T0284127

Kayaking the Sea Roads

Exploring the Scottish Highlands

Ed Ley-Wilson

Whittles Publishing

To my wife, Leah, whose care, patience, enthusiasm and ability
to suffer my ever-changing moods makes it all possible

Published by
Whittles Publishing Ltd.,
Dunbeath,
Caithness, KW6 6EG,
Scotland, UK
www.whittlespublishing.com

© 2023 Ed Ley-Wilson
ISBN 978-184995-563-8

Printed and bound in the UK
by Halstan Printing Group, Amersham.

Contents

Acknowledgements

A journey like this is never the construct of just one person and without the support of many lovely people, neither the journey nor this book might ever have made it this far.

First amongst equals of course is my wife Leah who understands me like no other and whose love and unwavering support underpins my abilities to head off into the unknown.

My publisher, Keith Whittles and his team, whose interest, enthusiasm and professionalism have been a joy to work with. I want to make special mention too of their editor, Caroline Petherick, whose expertise, humour, no small amount of patience and a highly honed critical eye helped steer the final text. Readers will be thankful for her red pen!

Huge thank you to my early readers: Corrin, Alec, Ken, Iain, Graeme, Tom and Dennis. Your early feedback and honest critique gave me the courage to pursue to publication.

Special thank you to my wonderful talented friend and neighbour Dot, whose beautiful illustrations help capture the movement, perspective and mood of this place that my own simple childish drawings could never do. Thank you too to Helen, another near neighbour, who has produced the gorgeous maps that help centre the journey in the reader's eye.

Throughout the trip, I was blessed both with the kindness of strangers and of old friends alike. The open welcome of the people of the Scottish Highlands and Islands is well known and I can confirm that it is indeed very real. Folk stepped out of their busy lives to carry, chat, feed, advise, give shelter and offer a weary traveller moral support just when it was needed most. Thank you Andy and Wei, the two Swiss ladies, Iain and Ben, Robert and Pauline, the Lochboisdale fisherman, the ladies in the Eriskay Community shop, Judy at the Applecross Inn, the Kiwi owners of Salen's café, the Calmac teams for their welcome and ready banter ... 'you're gonna need a bigger boat mate!', to friends Richard, Calum, Andrew and others whose quiet interest and support meant so much, and for the myriad smaller interactions on beaches and remote communities where a smile and friendly word so often helped to paint a grey day blue.

And finally, to those awesome people, past and present, who have inspired in me a hunger to take a peek over that horizon. To you: John, Alec, Ranulph, Dennis, Robert, Rog and others I owe a lifetime of restless enquiry and adventure.

Ed

Author's note

Many of the place names in this book are spelt in more than one way, so I have used the versions I'm most familiar with – that is, the ones in the now thoroughly tatty large-scale OS maps that have been so essential to my journey. In describing the wind, I have used the Beaufort scale and provided an explanation at the back of the book. I have also superimposed my own Red/Amber/Green system to provide an at a glance impression of the state of the wind and sea each day.

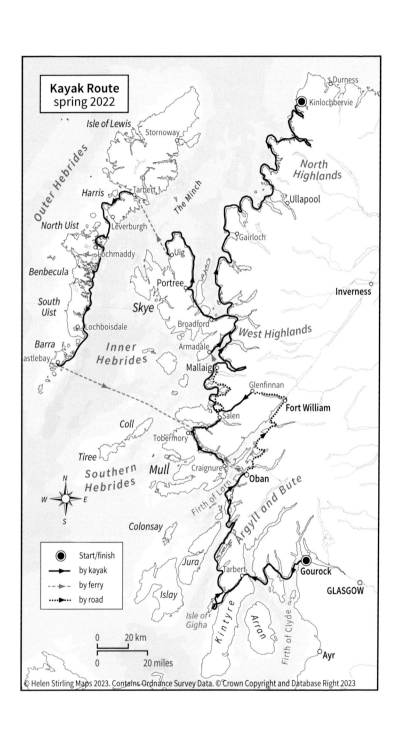

Kayak Route
spring 2022

Isle of Lewis

Stornoway

Outer Hebrides

Harris

Tarbert

The Minch

North Uist

Leverburgh

Uig

Lochmaddy

Benbecula

Portree

South
Uist

Lochboisdale

Skye

Broadford

Barra

Inner
Hebrides

Armadale

astlebay

Mallaig

Coll

Tobermory

Salen

Glenfinnan

Fort William

Tiree

Southern
Hebrides

Mull

Craignure

Oban

North
Highlands

Kinlochbervie

Durness

Ullapool

Gairloch

Inverness

West Highlands

Colonsay

Jura

Firth of Lorn

Argyll and Bute

Start/finish
by kayak
by ferry
by road

Islay

Tarbert

Kintyre

Isle of
Gigha

Arran

Firth of Clyde

Gourock

GLASGOW

Ayr

N
W   E
S

0   20 km

0   20 miles

© Helen Stirling Maps 2023. Contains Ordnance Survey Data. © Crown Copyright and Database Right 2023

Prologue: Feeling mortal

Archie, who had only two front teeth, was a trawler fisherman who couldn't tie knots. Because Archie couldn't tie knots, I was sat naked astride a throbbing outboard motor with my salmon farm workmates howling with laughter even as I saved their bacon. The boat had slipped Archie's rubbish knot and drifted away from the nets. None of the local boys could swim, of course – no money for swimming pools this far north in Assynt – so it was me who had to strip off and swim for the boat. Water so cold my undercarriage disappeared back up where it came from, arse exposed as I clambered over the transom, howls of laughter from the pens, and then that naked ride astride the motor, throttle arm suitably positioned between frozen thighs providing an impressive-looking phallus from afar, which was just as well because the real thing had joined his mates back up inside.

That day I was a story to be told, a butt of good humour and, according to the boys, 'a flippin' hero forever'. That day, too, I became the guy who didn't mind getting wet.

Thirty-two years later, with two months of solo sea kayaking ahead, I was to test that reputation to its limit.

I have never forgotten that swim for the salmon farm boat. Nowadays it would be instant dismissal, but back then – before mandatory lifejackets, Health and Safety reps and insurance protocols – I was the guy who would swim for a boat, and Archie was the guy who couldn't tie knots.

I loved that job and that life and those people. Salmon farming and making a life in the early 1990s up in the far north-west of Scotland was a chance for adventure and space and freedom. I sought it out with a hunger for the new and for the challenge, and found not only both of those but also so much more.

Fresh from military training, outdoor instructing, factory working and foreign expeditions, this was where I, along with my wife Leah, chose to lay our hats. We were young, probably a bit foolish, but full of passion and energy and a dash of devil-may-care. The passion led to Leah getting pregnant, the energy to us purchasing a small croft a mile and a half off any road and the devil-may-care to being economical with the truth to the bank about our income, which would halve as soon as we moved out into the wilds, to Kerracher.

But then, sometimes you have to do stuff that doesn't make sense, sometimes you've got to pursue the dream. You take the consequences too of course – in my case mostly financial it seems (I couldn't keep on top of the mortgage payments in the end) – but what is life without sometimes stepping into those marginal spaces?

Life at Kerracher, our tiny remote croft east of Drumbeg on the west coast of Sutherland, was dominated by the sea. I worked on the sea, salmon farming locally, and went home to a high spring tide that sometimes lapped at the bottom step of the tiny front garden, a boat so old that we had to bail it twice a day to stop it sinking on its mooring, and a long shallow rock-strewn beach, up which all messages, coal, guests, stuff etc. had to be carried. Winters were dark and long but full of adventures in windy seas. We would tie our son Thomas, too small for a lifejacket, onto a cleat, and to protect him from the spray we put a plastic bin bag over his head. It was the making of him, hardy adult that he has become.

In summer, I would run home over the hill from the fish farm and, still full of energy, set to work cutting, building, digging, constructing and repairing. There was always lots to do, a lifetime's work in fact, so prioritising tasks was key. We dragged out and dug in 2 kilometres of telephone line; kept chickens; ran the 'remotest B&B'; I wrote a book, *The Himalayan Shuffle*, about a recent mountain-running expedition; and we even built a mussel farm, so amateur and on the cheap that it broke up in the first autumn gale.

Ultimately, after a four-year struggle to make ends meet, my being economically truthful to the bank came home to roost and we couldn't keep up the mortgage. That and the fact that the local primary school at Drumbeg closed due to a drop in pupil numbers, meant it all became too much. We left in the mid-1990s, bowed but not beaten, and vowed we would find a way to return.

And return we did. Three years later, newly educated with a degree in anthropology and with a wee bit of cash in the bank, we were back – this time in Inverness, in a tiny rented flat and a more mature mindset, intent on making things work.

Twenty-five years on, the Highlands have been good to us. I've built a career in the salmon industry through processing and farming and facing squarely into retailers, NGOs, regulators and government; Leah founded and ran a Scottish recruitment business that was the making of us; we've experimented with buy-to-let; and we've engaged with community and charity in several ways. I have not had to swim naked for a boat again, but I have got wet a lot. The mountains and seas of the Scottish Highlands and ventures abroad have drawn my pals and me into all sorts of wonderful journeys. And sea kayaking has loomed largest of all. I love the freedom of it, the wildlife interaction, the infinity of choices. No rules – just your own skills, experience and judgement to carry you round that headland, through that lumpy sea, over that open space.

So here I am, less than two years short of the big sixty, in reasonable fettle and needing to take a break from business life for a bit. My reasons are simple enough. My work, though laden with values that support my worldview, is sedentary, and before I seize up into a chair-shaped ball of inflexibility I feel the need to move. But that's not the real reason. The real reason is that I've started to feel somewhat mortal. Over the last couple of years a number of my friends and colleagues have passed away or have developed life-changing conditions. Feeling mortal is making me ponder the 'what ifs' and reminding me that I have only one life to do stuff. I have things I *have* to do, such as work, of course, things I *need* to do, including family responsibilities, and things I *want* to do, like kayak, climb, write and more. The balance is a bit out of kilter at present, so it's time to take a risk and put myself back out into the empty spaces once more.

Taking to the sea by kayak is the journey of choice, no question. I have prior form of such expeditions in Norway and southern Patagonia a few years ago. In the latter my friend Robert and I had a scary and very wet expedition kayaking in and around the Magellan Straits exploring the windy and maze-like sea ways of the now extinct canoe cultures of the Alacaluf and Yamana peoples. It was a fabulous adventure, but we were limited by the time available, and that coloured our decision making. This time, in the Scottish Highlands, I am setting only a very rough timescale and keeping all options open. The freedom of that decision is just so exciting.

And I have decided to head west towards the seascapes of the Scottish Highlands and Islands. Why here and not abroad? Well, expense is one thing – there is a war on and inflation is playing havoc with my life savings – and I have done my share of running and walking and kayaking over distant landscapes and shores. This time I want to take time to think more deeply about what's happening here at my back door. The oft-called 'remote Atlantic Edge' is not so remote when you live here, and it is only an 'edge' from the perspective of those living far away, often in the urban centres of the south. The Highlands have a lot going on: farming, fishing, tourism, oil, green energy, forestry, land reform, community ownership, a thriving traditional music scene, devolved power, remote working, natural capital, world-leading access legislation, healthcare expertise, Scandinavian-inspired politics and economics – and of course, our so-called wild places. To kayak the physical landscape is to kayak the human one too, and there lies an opportunity to take some time to think.

In my journey through the Highlands, I am not going to be trying to break any records, I have no need to be 'first' or 'fastest' or 'bravest', as I feel I attempted all that conquering stuff in my younger years. This time I want to journey both outwardly and inwardly and to release myself to Mother Nature. I will be living outside, settling to new rhythms, and taking time to observe, to notice the small things, to listen, to absorb and to have space to think free from daily distraction.

It is a privilege, I know, but it will only be two months, and I have saved enough cash to do it.

So, just under two years beforehand, I set things in motion at my work to free me up without leaving any gaps at departure. It took well over a year to recruit new people, to bring certain projects to their conclusions and to navigate the ever-changing needs of our clients and leave them with a feeling of continuity of service.

That done, in June 2022 I found myself standing on the rough beach at Gourock on the mouth of the Clyde west of Glasgow, kayak packed to the gunnels and ready for the off. My lovely friends Rog and Anne and their son Andrew were there with digital bagpipes playing and a pack of jelly babies, which I was to take one of every time I thought I'd done well. My wife Leah was back at base, revelling in the newfound peace and quiet at home, but ready to support if I got stuck, injured, lost or waterlogged.

And what follows is that journey. Two months of solitude, of living outside and sea kayaking the west coast of the Scottish Highlands and Islands, from Gourock in the south to Kinlochbervie in the north. Just shy of 1,000 kilometres of solo paddling in what was to be the worst June/July weather the Highlands had experienced for many years. All the adventure anyone could want packed into an intense fifty-six days. Nerves frayed by seas too big for comfort, body hardened by the daily graft of paddling, and lumping kit and boat up and down rocky shores, mind opened by bedding into nature's rhythms, and soul enriched by encounters with life beyond our daily ken.

The flow of the narrative below keeps faith with the flow of the journey itself. I hope it gives you a sense of journeying and of sharing the daily challenges as they arise. Thoughts and ideas appear as they did during the trip, and are therefore in context with the physical environment I passed through. For those who like a bit of detail such as route, distance travelled, wind and sea state, I have included a summary at the foot of each day. These act as a sort of 'at a glance' impression of the day and include the occasional additional observation over and above what is within the main text. And finally, in any sea kayaking journey the wind looms large. Wind terminology can be confusing for non-seafarers so, at the back of the book, I have included an explanation of the Beaufort scale from a kayaker's point of view, and shone some light on how I used it to make decisions. I hope it helps.

Enjoy the trip.

1 | The quiet traveller

Days 1 to 7 – Gourock to Oban

Friday 3 June 2022, Day 1: Gourock to Ardyne Point, 21 km. Wind: Amber

Two years of preparation coming to fruition, and here I am at the end of my first day on a wee stony beach at Ardyne, 20 kilometres from my start point and looking out over the eastern channel of the Kyles of Bute. Sun-dappled water, light breeze, occasional sailing boat sails a-flapping, seals having a wee look-see, and the grass- and tree-covered hills of southern Argyll & Bute forming a 180-degree beach-eye view. Unbelievably stunning, unexpectedly warm, suitably remote – and as a result I have zero clothes on and have already had my first dip in the sea. Until late in the 20th century this place was used for military and other shipping ventures, and I can see the big engineering remains, concrete plinths and rusting chains dotted about. Overgrown and slowly being recovered by nature, somehow they don't spoil the beauty of the place, disappearing as they do into the vast surrounding landscape.

So this is Day 1 – Day 1 of a potential two months away from home. Leah has been wonderfully supportive, and after 35 years of marriage she knows me well and understands my needs. Having her support to enable me to disappear for all that time has made it all possible and avoided any negativity around the compromise of leaving too little time for the journey. Too little time for this sort of trip leaves your planning on tenterhooks all the time. Needing to get back by a fixed date limits decision making on both route and

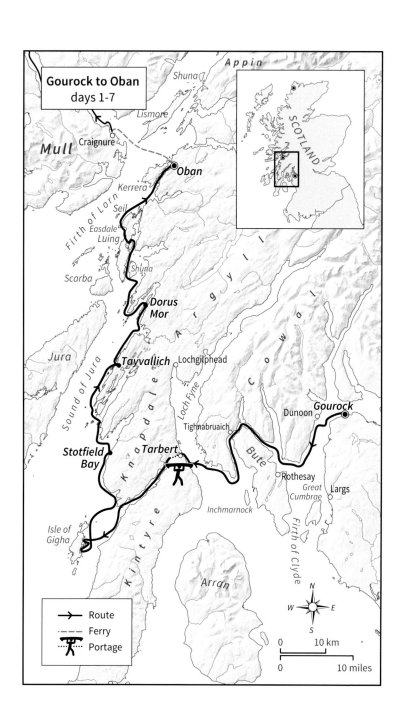

Gourock to Oban
days 1-7

Appin

Shuna

Lismore

Mull

Craignure

SCOTLAND

Oban

Kerrera

Firth of Lorn

Seil

Easdale
Luing

Shuna

Scarba

A r g y l l

Dorus
Mor

Jura

Sound of Jura

Tayvallich

Lochgilphead

Loch Fyne

C o w a l

Gourock

Dunoon

Tighnabruaich

Bute

Stotfield
Bay

K n a p d a l e

Tarbert

Rothesay

Great
Cumbrae

Largs

Inchmarnock

Firth of Clyde

Isle of
Gigha

K i n t y r e

Arran

N
W   E
S

Route

Ferry

Portage

0   10 km

0   10 miles

distance right from the start, and having the luxury of a two-month period means I can relax.

I won't relax, of course! Well, I will, but not in a traditional sense. I am going on a journey. It will appear as a physical one, but the real journey is inside. My original plans to paddle to Norway with my friend Angus were scuppered earlier this year when he developed carpal tunnel syndrome in both hands. That and a financial crisis and a war in Ukraine made the expense of a Norway trip a bit bonkers. With operations on his hands coming second to covid patients in the hospital, there was no chance of Angus being mended in time for today's 'off'. So plans change. It's what happens.

My trip now is to spend time kayaking the west coast of the Scottish Highlands and Islands, roughly south to north. No record attempts, Facebook or blogging – just me, the sea, the islands and lochs, the wildlife, the weather and the open space, physical and mental, over my bow. Am I anxious? Yes. Excited? Yes. Feeling a bit selfish? Yes and no – not really, to be honest. It's only two months, and I have done enough focusing on others over the past months to take a wee bit of time out for myself. In my head are Jim Hunter, Alastair McIntosh, the Canoe boys, John Muir, Thoreau, Boswell and Johnson, Frank Rennie, George Monbiot, Norman MacCaig and many more. Blessed as I am with years of reading and an ongoing love affair with the Highlands, this place through which I am going to travel is more, much more, than just a view. Living and working here for almost 40 years, I see something of interest everywhere. Ecology, biology, economics, politics, land ownership and social life. Like the geologist who sees not just the rock in front of them but the very act of creation, the anthropologist in me is keen to take a deeper dive. I remain fascinated, both to understand the ecological and social issues as they appear, and hopefully to contribute some thoughts on changes needed. Ultimately I want to ensure these Highlands of ours remain not just as picture postcards for tourists, but living, breathing, working spaces for their people, with community and environmental care as core.

So, today. Today is the 'off'. I am not going to give chapter and verse to every move of every day but will instead précis where I can, and take a deeper dive as and when the mood takes me.

The day is not unchallenging. My friends Rog, Anne and their son Andrew are wonderful. Rog, who undertook a similar journey over 25 years ago, understands what it's like to head for the horizon on your own, and he is sensitive to my poorly hidden nervousness. He and his family have kindly driven me to Gourock, my starting point, where the River Clyde leaves the land and heads out to sea. I wanted to start on the Clyde in recognition that our West Coast, rural as it is, is intimately linked with our urban centres – and especially, historically, with Glasgow. Goods, services and government agency would all travel north from Glasgow, first via sail and then via steam as the puffers, small steam-engine boats, became a lifeline supporting remote island life.

Now I'm out from the jetty and into a grey sea on a grey day in what feels like a bit of a grey town. And then a dolphin, a bottlenose, surfaces in a high arc right in front of my boat. Muscular shades of glistening grey and a good 2 metres or so long, it exudes power, and I catch my breath. I am only 15 minutes out from the shore, anxious about the journey ahead – and then this wonderful animal comes to say Hi. It feels like a visitation, and my heart leaps at this blessing from the deep. I know that's floppy thinking, but I'm happy to run with the emotion of it and accept such feelings as they come without needing to reason everything out all the time.

Weaving my way through the moored boats on the western end of Gourock, I look for the dolphin again and catch just a fleeting glimpse of a dorsal fin heading away upriver – and then it's gone. As I push out into the crossing to the old Victorian resort of Dunoon, other animals come to visit: eiders with their saucy 'owooos'; small rafts of greylag geese with goslings in tow; piping oystercatchers; and nervous-looking shag – the latter, an ancient survivor for many millions of years, looking shifty as ever, like they feel guilty for being around for so long.

And there are boats galore, too, at this gateway to the sea. Big ones, little ones, ferries, yachts, cargo vessels, all criss-crossing the firth between Gourock, Loch Long and Gare Loch. They leave wakes of all sizes, and I bob or lurch over them as they come my way. The paddle steamer *Waverley* comes by, long and narrow, and with her twin red, black and white funnels characteristically swept to stern. She looks regal, neat, a design-treat for the eyes.

Crossing to Dunoon, I find wind and tide lump up the sea into a bit of confusion out in the middle; then a couple of hours later, down at Toward Point, where I turn west, there is some light surf that adds further frisson to the day's paddle. The Isle of Arran looms large way off to the south and the Isle of Bute to the west, and before long I am paddling into a clearing sky and a calming blue sea as I head to the shore at Ardyne Point for my first camp.

Back at Gourock, Rog and Anne had kidded that travelling alone I would soon be talking to myself, going a little crazy perhaps. What I needed, they'd told me, was a 'Wilson', Tom Hanks' bloodied football face in the film *Castaway*. And now, I kid you not, only 20 kilometres from the start, as I am carrying boat and kit up the stony beach at Ardyne, there is Wilson peering up at me; a small yellow rubber ball with a printed smiling emoji-style face tilted to the sky lies partly buried and covered with seaweed. I laugh out loud, brush the seaweed off him and sit him on the edge of my deck bag, where his wee smiling face will greet me each morning. And yes, it is a him. I don't think my wife would appreciate me spending two months away with another woman. Day 1, and already I have a companion.

Watched closely by my ever-smiling new pal, I pitch the tent on the rough lumpy grass above a stony beach, strip off and swim, revelling in the delight of a sun-kissed late afternoon and a first day done. Shoes off, brew on, and now disciplining myself to write up the journal. From the mouth of my tent I have a view straight up the east entrance of the Kyles of Bute, and as I look up from my writing a sea trout has jumped clear of the water just a few metres offshore and three porpoises are hunting further out. What a window!

- · Route: Gourock to Ardyne Point
- · Distance: 21 km
- · Weather: grey and a bit breezy until end of the day when the sun appears
- · Wind: Amber. Force 3 gusting 4; wind on the stern and then on the beam, down to Toward Point
- · Sea state: generally lumpy, and calm after Toward Point; crossing to Dunoon is only 3 km, but wind and tide lift the middle section into a confused bit of water

- Events: dolphin visitation, swim, hunting porpoises
- Camping: rough ground, but gorgeous view west into the Kyles of Bute.

Saturday 4 June, Day 2: Ardyne Point to Port Leathan (east side of Loch Fyne), 29 km. Wind: Green.

Last night was picture-perfect. Sun dropping slow to the horizon, sea all a-glitter, sea trout jumping, porpoises hunting, gannets diving. Crawling out of the tent in the wee small hours, I found the night warm and dark enough for some stars and a moon reflecting a rippled sunlight on the water.

Gazing up into that night sky is one of life's great pleasures. It is almost midsummer, and half a million kilometres away a crescent moon is reflecting enough sunlight to crowd out many of the stars. But still there are plenty to wonder at, and looking up is to take a journey back in time – distances and timescales so vast that some of those distant suns may already be no more, extinguished years ago with the light of their final dying days still to travel our way.

And there is the moon, suddenly playing a major part in my life for the next couple of months. Its gravitational pull will colour my every day. Its tides will govern how and when I can move, the speed of flows through narrows, the distance I will have to carry kit up off the shore, the size of waves where the flow moves against the wind. Tide and wind. These are the forces that connect my tiny person to the sky above. …

I crawled back into the tent hopeful for many starlit nights to come.

With the day dawning bright, I take time packing and carrying kit and boat up and down the beach. No point in suffering an injury this early on – indeed no point in getting injured at any time, of course, but I need to set a behavioural precedent by moving carefully when shifting the heavy weights of kayak and kit. I'm in this for the long haul. I'm also starting this trip with a dodgy left shoulder, which I'm hoping the constant movement of daily paddling will ease. Overdoing the carries would definitely do it damage.

Check Wilson is okay – he's still smiling – and cross to Ardmaleish Point on a flat blue sea then paddle a slack tide up towards the Burnt

Islands at the northern end of Bute. A few other early risers are about, mostly yachties, and some give a wave. Gorgeous-looking properties on the mainland shore and, tucked into the trees, a delightful wee church with a spire. I see the ferry, the *Dunvegan*, on its early morning crossing up ahead, and wait for it to pass before moving north. Rest at northern tip of Bute just shy of Buttock Point (strange but true). Yoga in the sun on a sheep-spattered shore to help ease out the shoulder. Observe some tidal flush but nothing of note really, so paddle on against the tide on the west side of the kyles, heading south. No problems with racing tidal flows in the end – the kyles are wide enough to slow the spring movement – and I am fit enough, it seems, to paddle against it. Gorgeous going past Tighnabruaich, and I spot Canada geese and whooper swans. I say 'spot', but in fact I'm right on top of them before I know it. Goslings a-plenty, fluffy and flightless, dash about in some panic as I appear, so I don't linger.

South, then, to round Ardlamont Point, having stopped for a bite to eat and more yoga on a pebbly beach. At the bottom end of the kyles the world opens up with stunning views of Arran's hills and southern Bute looking green, arable and just lovely in the sunshine. A final paddle with open water to the south, and on to the day's end at Port Leathan, not an actual port but a picturesque wee cut in the rocky shore with a bite of grass for pitching the tent.

Another sun-kissed day. Body tired, but I have an inkling that my injured shoulder might be okay going forward. In this good weather it's too easy to become complacent and overdo the distance or the effort. It's only Day 2, and if I pay attention to the yoga and careful carries of kit then I'm hoping my body will readjust to this way of living.

A couple on a yacht let me know the forecast ahead, and it does not look so good. According to the long range forecast this fine weather is due to break into what might become a prolonged spell of wind and rain. Time to plan ahead. I have decisions to make. The most pressing is whether to attempt to round the Mull of Kintyre. That would mean several days of paddling south, then round it and back up the west coast of the peninsula. The distances and forecast suggest I would likely get round in the last of the clear weather, maybe – but I would then get pinned to the western shore, perhaps for days if the forecast proves correct. The conqueror in me would love to round the point, but I remind myself that that's not what this trip is about. I'm here to travel within, to bed in, and not to stamp myself upon the landscape. More Nan Shepherd than Doug Scott. I've done my share of conquering peninsulas and mountains, so now, alone here in the west, this is my chance to settle to a new rhythm. I am determined to learn to move more slowly, to think more deeply, and as a result perhaps come to better understand this place through which I am travelling and from which I have derived so much joy for so long.

So the forecast, the time available and the trip's value system combine to let me know that I will need to keep travelling west and then north to get myself to a point where, when the weather breaks, I will have options.

- Route: Ardyne Point to Port Leathan at mouth of Loch Fyne
- Distance: 28 km
- Weather: sunny
- Wind: Green. Light, 2 to 3, some on the nose, some behind
- Sea state: slight
- Events: wildlife bonanza. Porpoises, seals, herons, Canada geese, whooper swan, buzzards, birdsong in the trees, butterflies, wacky looking bugs (I really need to improve my entomology), bees, cuckoo, roe deer; meet a couple moored up in their yacht, both in their eighties and retired for twenty-seven years and living life on the water
- Camping: private spot up off remote rocky beach, perfect. Sitting here writing, totally starkers, and soaking up the sun.

Sunday 5 June, Day 3: Port Leathan to Gigha, 29 km. Wind: Green

What a day … I'm lying in my tent on Gigha! Never expected to make it here today, but weather and luck have been on my side and the body felt okay so I went for it. A day full of contrasts.

Up early and pack, breakfast and away by 07.30. Only a 3.5-kilometre crossing across the mouth of Loch Fyne but the wind is up, and a rolling sea is pitching up on my port beam. Spray on the face is a surprisingly good wake-up call, and the moment I leave the headland the wind is on and it's time to concentrate and knuckle in. The feeling of exposure is very great, and travelling alone changes everything. It is a mind game I am slowly dealing with. This early in the trip it is still a nervous 40 minutes or so of bouncing around on a lumpy sea, totally alone, early morning, and no chance of baling out if the wind and sea rise another notch. In the middle of the crossing the land feels a long way away. But Wilson is still smiling, the sun is coming, the sea is turning blue and white-flecked with occasional breaking waves, and a surging mix of nerves and elation floods my senses.

I paddle hard, in open-crossing mode at around 8 kilometres per hour to get across the open space, then turn right once I come close to Loch Fyne's western shore. Wind and sea now loom behind as I surf slowly north towards Tarbert, my gateway to the west.

The shore south of Tarbert is strewn with rhododendrons. At this time of year they look lovely, but these are an introduced species, *rhododendron ponticum*, and they appear to be slowly taking over our hills and glens. As they grow into thickets, little if anything can grow beneath them and so wide areas of monoculture develop. If it were up to me, I'd stem-inject the lot of them to kill them off and allow our native species to poke their heads up into the sun again. Some rhododendrons would survive, but would do so as part of a diverse floral landscape rather than the monocultural deserts we see at present. Planted by our Victorian forebears both for privacy and for their flowers, these massive shrubs have been left unchecked for far too long, and we need to do something about them. Until recently, I have been chairing a small charity, Coille Alba, run by a friend of mine, biodiversity expert John Parrott. John's work focused both on the control of invasive species and the preservation of native ones. Rhododendrons were on our hit list, and our small team of field workers have done great service in removing large areas of them from various parts of the Highlands. You need landowners' permission, of course, and not every landowner is ecologically aware enough to care until it is too late. There is much to do.

There is a debate to be had here, of course. When does an 'introduced' species become 'invasive'? Should we treat introduced animals differently from introduced plants? Is our vision of a pre-introduced landscape realistic or meaningful? And who decides, anyway, what species our landscapes should contain? Author Emma Marris explores some of these themes in her book, *Wild Souls*, and makes the point that when we prioritise one species' existence over another, there are some ethics to tackle. The ethics of removing animal species centres around the culling of individual animals, individual souls. The physical culling of individuals involves suffering – it is a messy business – and when they, as an introduced species, are not themselves guilty of invading, when they're just doing what's natural to them, there is a lingering question around whether it is ethically robust to favour them over the natives. By deciding to cull the black rat from a Scottish island, the possum from New Zealand, the cane toad from, well, just about everywhere, we are making a choice about individual rights. We are deciding ultimately that one animal has more rights to exist than another. We are exercising preference.

This issue is easier to handle when dealing with plants, I think. Of course, we could debate too whether plants, which do make decisions of a sort – for instance to turn to the light, or to engage with other species, such as fungi in symbiotic relationships – have agency of their own. But to me 'agency' suggests conscious intent, and that is where I start to struggle with the idea of aligning plants and animals in the ethical debate around suffering.

And as for deciding what species our landscapes should contain – well, favouring species is what humans have always done. The moment we started chopping down trees for houses, or burning undergrowth to flush out prey, or planting crops, we were deciding what sort of landscape we preferred. We did it then and we do it now, and there is a discussion to be had here around what part our cultural baggage plays in determining how we think about our place in nature. I'll explore that theme further later in the trip. For now, though, gazing up at this blanket thicket of rhododendrons, I feel comfortable enough about wanting to get in amongst them with a stem-injection gun. My own favoured landscape, in this case, is for floral species left over from a post-ice age, pre-human existence: birch, hazel, rowan, aspen, oak, alder, pine, juniper and more, perhaps with a little rhododendron peeking up in a clearing or two. That would be fine.

/ / / / /

As the shore begins to bend west in towards Tarbert, a yacht and then another yacht and then a fleet of yachts in succession come motoring out of the mouth of the loch and set their sails to the wind. As I paddle past and put ashore on a small pebble beach, a wonderful chaotic and informal regatta is taking shape out in the open water. Early-morning sun glinting off white sails and hulls, boats of all sizes tacking and gybing and somehow not crashing into each other. This is the Jubilee celebratory regatta. Not a competition, just a colourful, reckless devil-may-care dashing about of yachties celebrating their Queen's seventieth year of reign. These sailors look like they're having fun and it's a testament to their collective skill that there are no collisions. A beautiful sight out there in the sunshine and stiff breeze.

The small picturesque village of Tarbert straddles East Loch Tarbert Bay and, as its Gaelic name describes, it sits at a crossing point, a narrowing of the land between bits of sea. Here, it is only a mile or so to the west side of the Kintyre Peninsula, and my task now is to get boat and kit across. The sole Tarbert taxi man is not answering his phone, and Dave from the Tarbert-based kayak guiding company is away with one of his courses. I therefore roam the street (there is only one, really) spying out vehicles with roof racks, with a view to asking for help from their owners. It's a long shot and a bit cheeky, but you gotta do what you gotta do. Nothing doing for an hour or so, and then I spot Andy and Wei in their battered van with the all-important roof rack. At that point I don't know they're called Andy and Wei, of course but after I have gently knocked on the window and put my request, Andy doesn't even flinch. Wonderful people, and their attitude to helping is the epitome of the kindness of strangers. Andy from Portsmouth and his partner Wei from China, adventurers both, free spirits currently travelling about the country, take my kit into their van and my boat onto their roof and drive, gears screaming, the mile and half to the fishermen's pier on the west side of Tarbert. We have some great chat, and although they don't want anything in return, I manage to squirrel a 'thank you' under the cushions in their van; I just hope they find it at some point. There are many ways of being in the world, are there not? – and, while Andy and Wei were happy to communicate and help, the guy parked behind them in Tarbert, sitting grumpy-looking at the wheel, had a sign hanging under his mirror saying, 'Piss off, I'm having a bad day'.

Long paddles west down the ever-widening loch follow. A stop halfway for some stretching, the wind rising behind as I near the seaward end, a white beach to rest again, a small boy running over to jump into my kayak where it sits on the sand and drive it like a car, then a long pull south and a curling coastline to get as far west as possible before crossing to Gigha. And then the crossing. I don't know whether I'm going to make a go of it until I get to Rhuanahaorine Point at around 7pm, and the sun is sitting ever lower in the west. The sea round the point looks a bit busy with wind, but the temptation of getting to Gigha is too much, and with the ever-present threat of a weather change just a couple of days ahead, I think I should press on.

Out past the headland, everything changes again. Leaving the relative calm of the mainland to stern, bow pointed west, I paddle into the setting sun and a windy sea. Out in the middle, each shore is only a kilometre and a half away but, rolling over the waves with the sun in my eyes and the spray on my face I feel exposed – yes – but in control. A wonderful feeling of elation comes over me, the boat riding the waves, my paddle dipping and rising, the sea bowling in on my beam. I'm in the element I love, immersed in the moment, journeying alone. At one, if you like. That phrase can seem a bit corny, but out there in the moment, in the risk, one's place in the world is brought into sharp focus and emotions run high. I am feeling as well as thinking my way. The latter is always fascinating and never-ending, while the former is more of the moment. A flush of chemical to the brain, an enculturated interpretation, an instinct born of wonder, joy or fear. Accepting that feeling when it comes, without overthinking it, can be a wonderful thing.

Details of the far shore start to emerge from the previously distant smudge. A rocky face here, a stepped hillside there, individual trees, houses, all take shape. Then I am there at last, alongside the east coast of Gigha, turning right to round yet another headland and take shelter west of Ardminish Point. The sun is low now, and blinding to look at, but the tide is high and getting ashore is not a problem. As the bow of my boat crunches on to the shell-strewn rough sandy shore, curlews, oystercatchers, redshank, greylags and ringed plover set up a chorus of alarm. There is goose poo everywhere, but I don't care – I've arrived in the west at last. The birds settle as I pitch my tent just a couple of feet above the high tide mark, and they go silent once I disappear inside. The wind dies, the midges rise, I cook, then zip up, eat and sleep with the quiet satisfaction of a day well spent.

- Route: Port Leathan to Tarbert; vehicle portage to West Loch Tarbert. Paddled to Gigha and camped in western end of Port nam Faochag
- Distance: 29 km
- Weather: sunny, warm, blowy
- Wind: Green. East-south-east all day, Force 3, on the beam and stern
- Sea state: mixed; lumpy crossings, but West Loch Tarbert to Rhuanahaorine Point had the wind on or around my stern most of the way
- Events: memorable crossings, regatta and ladies swimming in Tarbert, Andy and Wei's kindness, yoga on the point halfway down West Loch Tarbert. Swims at Dunskeig Bay and at the mouth of West Loch Tarbert. Greeted by birdsong in Gigha. Midges!
- Camping: Arrive at high tide, knackered. Gorgeous view across the sea loch out of the tent entrance.

Monday 6 June, Day 4: Gigha to Stotfield Bay (Knapdale), 25 km. Wind: Green.

First real midgy night last night. Occasional breeze, but any pee-venture without protection was doomed to a death by a thousand bites. The mozzie coil came into its own once again. Any breeze died around 9pm and I awoke to a calm sea.

Breakfast of midge-infused porridge and a steady packing of the boat on a wet slippery shore on a dropping tide. Back feels a bit sore today, and shoulder is tender again. Am moving slowly, taking the long view and trying to stay off the drugs even with the Ibuprofen winking at me from the first aid kit.

Paddle round to the village and put ashore beneath The Boathouse restaurant on a tiny slab of white sand. Exquisite location, but the restaurant is closed that early in the morning. Take a walk up in the early sunshine through Ardminish village and out to the gardens surrounding Achamore House. Gardens are not usually my thing, to be honest, and I don't know what to expect. My attitude has now changed. What has been achieved here on Gigha is quite astonishing, and the effort, planning and obvious deep knowledge and passion are all there to see. In summer Gigha looks and feels lush; studded with walled fields, rough ground and uplands, it is today a green jewel set in a turquoise sea. But look closer and you will find another Gigha: a managed garden dating back over 100 years, with flora from all over the world and, at this time of year, flourishing with colour and new growth. A private place but open to the public, and taking its place alongside any mainland National Trust property for sure.

Look closer still at Gigha, and you will discover an island and its people working hard at taking responsibility for their own futures. Relatively sparsely populated, the island was one of the early adopters of community ownership, and in order to make ends meet and hopefully to flourish there is innovation everywhere you look. There are wind and salmon farms, and it is clear that many of the houses have been newly insulated, too. The place looks well kempt, not posh or managed, but lived in. This is a far cry at present from other parts of the Highlands where incoming money is building/renovating ever bigger and fancier holiday homes or, with some of the fiercer rewilding projects, keeping people off the land altogether. The holiday home/second home piece is especially damaging, as local people and incoming professionals such as teachers and nurses are priced out of finding somewhere to live. I spoke recently with a salmon farm manager on another Hebridean island who was clear that he was destined to live with his growing family in rented accommodation for ever, priced out of the house purchase market by the big money from the big cities. Second homeowners, alas – they know not what they do, for in buying their two-weeks-a-year slice of heaven they are killing community as we know it.

What lies in wait, if we're not careful, is Toy Town, Venice, Padstow, Benidorm. Well, perhaps not Benidorm – the sea is too chilly here and the midges too numerous – but the Highland communities I have spent a lifetime living within and which I would like to see continue are those which are predominantly permanently occupied, vested and with a strong cultural sense of place. Such communities are not always perfect, and there are comings and goings, but they are at least full of folk committed to making a life in that place. Commitment, not commuting – that's what makes a community tick.

Back at the beach, a quick feed from my snack-bag, and away. The windy forecast ahead is starting to loom large. I decide to take a longer open crossing, so paddle up to the north end of the island where there is a wonderful seaward feel with eiders cooing and a pair of divers quacking overhead. I then turn north-east for Rubha Cruitiridh on the mainland, about 8 kilometres away. From this distance in this light the mainland shore is without features, a low-lying distant smudge of blueish colour, but the sea is flat calm so I set off without concern.

A couple of kilometres out, the only movements on the water are the swirl of my paddle, a light wake astern and the occasional cormorant or seagull paddling slowly in the heat. The only sounds are a light splash as my blade dips, and distant seabirds' cries. I am 'alone, alone, all, all alone, alone on a wide wide sea'. All is quiet, all is well, except of course when it isn't …

Far off to the east, a small white blob appears out of the mouth of West Loch Tarbert. Oh shoot, it's the ferry! I'm on a ferry lane – right on it, actually – and as the blob morphs into its ferry shape, it turns towards me. Now I can see the white of the bow wave and start to hear the quiet throb of its engine. I am paddling north-east, directly across its path, and I start to worry about collision. Well, it would hardly be a collision, would it? – more a flattening. I know from previous experience that kayaks can be almost invisible, especially from afar, and especially from a high point such as a ship's bridge. As I paddle north-east the ferry's bow keeps turning to point directly at me. That's collision course. I need to see either its port or starboard side to know that I'm not going to end up underneath the thing – but the bow, it just keeps pointing right at me. What to do? The calm of the sea is no longer mirrored by my own demeanour. I consider radioing the skipper and turning on my strobe safety light, but the distance between us is diminishing rapidly. Instead, I put my paddle in the air and wave it about a bit, hoping the white flash of the blade will attract the skipper's attention, and then set to paddling hard. Any experienced kayaker will have three types of strokes: pootling along all day (6 kilometres per hour); crossing an open space in weather (8+ kilometres per hour); and then a 'get the f..k out of here' stroke used to, well, get the f..k out of there! Keen not to get run over, I engage the latter. Ten minutes of sweat-flying paddling later, and the starboard side of the ferry slowly starts to come into view. By then it's only about a half-kilometre away and closing fast, and I can now see the wake it's leaving, too. I have no idea if they have seen me or not, but as the ship passes behind me at speed I'm clear by only around a couple of hundred metres. Nerves jangling, I ride the big lazy wake easy enough – and then need to pee. One crisis replaced by another. Thank goodness for the onboard bailing jug!

Reaching the mainland shore at Rubha Cruitiridh, my body is tired, and I put in for rest at a seaweedy beach. Limestone cliffs, caves and hidden coves to landward, a peregrine falcon hunting, its call echoing off the cliffscape, and my first otter of the trip. Common seals, reduced to blackened bobbing heads, come curious, and I rest in the sunshine. Very tempted to camp here, but the urge to travel is strong and Stotfield Bay, 10 kilometres to the north, is winking at me from the map.

So that is where I am now, camped up, fed, full body wash in a small stream, baseline clothing drying (hopefully) on the pebble beach, and a small fire crackling away at my feet. The Paps of Jura, 20 kilometres away across the sound, are blue turning pale pink. In the past I have paddled round Jura and part of Islay, and at one point I thought I might go there during this trip too, but now I know that the forecast ahead would leave me stuck there for I don't know how long. And anyway, Stotfield Bay is in a part of the Scottish west coast I have never visited, and the excitement of exploring new ground is kind of addictive.

- Route: South to Gigha village, then north to northern tip of Gigha and across to Rubha Cruitiridh then north again to Stotfield Bay
- Distance: 25 km

- Weather: cloud, weak sun, warm
- Wind: Green. None!
- Sea state: flat calm
- Events: Gigha gardens, lushness of the island, *rosa rugosa* hedges (memories of Kerracher), exposed crossing, ferry, otter, peregrine, shelduck, eiders, shag colony, fishermen hauling pots
- Camping: White beach with loads seaweed … messy and bit smelly. Easy grassy slope, good campfire, views over Jura.

Tuesday 7 June, Day 5: Stotfield Bay (Knapdale) to Craignish, 37 km. Wind: Green.

Feels like a big day today. Calm seas again – a rare spell of such weather here in the Highlands – but bigger winds are due from tomorrow, so I'll need to put myself in a position where I'm not trapped for days on a windward shore. In a perfect world the weather would permit me to move further west and explore the islands' western edges, but that is not my reality. Winds are rising to Force 7 on Saturday and Sunday, and moving to the south-west. The sea will build and build, and for a man in a small boat, moving with body power and no engine power, I must cut my cloth. Decisions on today's route have taken all that, plus the geography of course, into account. As a result, I push on and am a little further north than I'd expected to be at this stage of the trip. The alternative would have been to head west, just because it is, well, west – and then get pinned.

This part of the coast is gorgeous. Not grand or imposing, but indented with sea lochs, pointed headlands and islands offering a seaward feel. Birdlife is prolific, with herons, goosanders, Canada and greylag geese, islands of terns, sand martins, buzzards, mute swans and gulls, all lazing about in the quiet air. Otters too have put in an appearance. Never lazy, always busy, almost frenetic really, they must eat a sizeable percentage of their body weight every day just to stay warm.

North-west of the mouth of Loch Sween I put ashore at the old pier at Rubha Riabhag. The water is green with weed at low tide, and the stacked stones of the pier are painted with sea pinks. Far off, the Paps and the Jura coastline haze a pale blue above a mirror-calm sea. A swim and a quiet yoga in the sun while drying off. Life is good.

Moving on north, at Carsaig Bay I walk over the peninsula to the small fishing village of Tayvallich in the hope of refuelling both food and mobile phone. I need the latter primarily for forecasts, and I have resisted using it to communicate daily with friends and family. Everyone knows about the trip, and all have been respectful of my desire for some solitude. Pub closed, coffee shop closed, but wee shop open, and the helpful young man in the latter agreed – for £5 (!) – to put my phone on charge and sell me some almost-off leeks and carrots. Beggars can't be choosers.

Back on the water, newly revictualled and batteries energised, I bypass Crinan, where boats transiting the Crinan Canal exit into the Sound of Jura. Leaving it a couple of kilometres to the east, I strike out from Ardnoe Point for Loch Craignish; landing is difficult with greasy rocks to negotiate on a low tide and a long carry of boat and kit up to the grass. Spend some time toing and froing between possible landing spots, clambering out of the boat, slip-sliding up the seaweed to scout possible campsites, finding nothing, back down to the boat, on to the next spot and repeat. Makes for a testing end of day. Finally get the tent pitched at 8pm in a remote spot on the north shore of the loch; I am now ensconced there, and have just downed my now-regular rice and tuna dinner.

Behind me and over the hill lie the ancient prehistoric cairns and standing stones of the Kilmartin Glen. Raised before the pyramids of Egypt, these huge rocky mounds of chambered burial cairns speak of societies and belief systems long gone. It is a landscape rich in ritual meaning whose details are lost to us now, but which are ripe for conjecture and a vivid imagination. It was from here, on the shores of Craignish, that those ancient peoples may have taken access for fishing, raiding, trading, scavenging, socialising and no doubt escaping from attack, too – my imagination runs riot: deerskin-clad hunters paddling log canoes, or wooden-framed coracles or Irish-style currachs, heading out with grass-wound twine and fishing hooks of sharpened bone or antler to fish for their supper. I imagine a sea still heaving with fish, huge pollock and skate pulled to the surface to be speared and heaved on board. Nature's bounty, right here, just

Early days through the Kyles of Bute

Blowy crossing to Gigha

Calm at Stotfield bay

off from where I'm camped. Did these ancient folk, like the Native Americans, then give thanks to some spirit of the sea? Was there reverence? Did they cast offerings to appease supernatural powers or to secure ongoing plenty? We can't ever really know, but sitting here on those same shores, pinned in my tent by the midges and nervous about the rising winds I can hear rushing through the trees above me, I welcome the thought that my own experience on these seas (Gore-Tex jacket and neoprene trousers aside) is perhaps not so far removed from those of long ago.

I am very tired now, and tomorrow I have the Dorus Mòr to deal with.

- Route: Stotfield Bay (Knapdale) to bay at Creagan nam Buachillean to mouth of Loch Craignish
- Distance: 37 km
- Weather: cloudy start, then sun
- Wind: Green. Slight and some calm
- Sea state: mostly calm
- Events: birdlife, yoga in the sun, weather starting to change
- Camping: difficult access, just enough space for my tent once I had cleared the ground. View south to Jura and back to Knapdale. I'm in a good position to make decisions as the weather deteriorates going forward.

Wednesday 8 June, Day 6: Craignish to Luing, 20 km. Wind: Green/Amber.

Two points of interest today: passing through the tidal rapids of Dorus Mòr, and the stunning camping I find at the top end of Luing.

The weather has broken as forecast. Last night it poured with rain, drumming away on the tent fabric just inches above my face. Morning came and still the rain poured. My forecast has suggested it will clear around midday so rather than get all my kit soaked in packing up, I spend the morning in the tent, resting, stretching and planning ahead. The winds are due to follow the rain and I have no wish to be pinned, so a bit of careful planning is key. Ahead, just a few kilometres away, is the Dorus Mòr, my gateway to the north. *Dorus mòr* translates from the Gaelic as 'big door' and it's exactly that, an 800-metre gap between the mainland and Garbh Reisa, the island to the south. Millions of gallons of water are squeezed through here on every tide. I have been through before, many years ago and I have sat at Craignish Point too and watched wind against tide throw the sea into a confused surf. Here, on my own, I need to get this right. If I capsized there's no one here to pull me out of the sea, and the tidal race through the Dorus is so strong that I'd be a kilometre out from any land and on my way towards the Corryvreckan whirlpool before I got back into my boat.

So I do the calculations to work out slack tide as best I can, manage to confuse myself, so look outside instead and suss the tide as it's happening in front of my nose. I know we're in a neaps tidal pattern at present, when the level of tidal change is less than at springs. I know too that high tide today will be around 12.30, so that's enough. I will aim to enter the Dorus Mòr around 12.30, expecting some movement but hopefully not so much that I will need to retreat or indeed get sucked into a flow too strong to paddle against. This is one of the things I love about kayaking. A rough and ready reckoning will do most of the time. You've just got to look outside. If you know your rough times for high and low tides and you can suss springs or neaps and you know the general direction of the incoming or outgoing tide, then with the wind speed and direction in mind you can usually make a fairly good guess at the conditions ahead. In a yacht it's different, of course. For one, there's a lot more money involved if you screw up, and two, you can't put ashore if it all goes wrong. And then there is the maths and the rules and the lights and the speed over the ground and much more. If I had a yacht – and I am not in that financial bracket unfortunately – I suspect I'd be worrying so much about the cost of crashing that I'd get the maths wrong and crash anyway.

Towards midday I pack up, still in the rain, and set off. Wind from the east, freshening and a lumpy gusty bit of sea heading to Craignish Point. As it turns out, the approach to Dorus Mòr is more gnarly than the rounding itself. As I round the first of the two headlands the water is still moving west and north, and I get out and run ahead to peek at the flows lying in wait. Back in the boat, in the flow, a couple of eddy lines snatch at my boat but other than that it's easy enough to get round and start heading north. Way off to the west, the Gulf of Corryvreckan looms large, and just offshore from my passage the eddy lines are standing up and the sound of rushing water is close by. It's quite a place; a gateway and a graveyard for many travellers over the years, this upper section of the Sound of Jura is a pinch point. Tides running north must squeeze between small gaps in islands and, coupled with a complex underwater topography, the water rushes and boils and stands up and falls and whirls. A paddler must pick their way through this ever-changing liquid landscape, and there's no time for complacency, for sure.

Early travellers to the Highlands from the Scottish Lowlands or further afield, back as far as the 17th century, will all have passed this way, back then journeying as they must by boat. No cars to whisk you effortlessly through the landscape at speed, no planes to leap over mountain barriers, and few tracks to ease your journey by foot or by horse. Most of the coast would have been a trackless place, hard won by those with the need or desire to explore. So everyone went by boat – which is great until you get to the Corryvreckan! Here, the gulf sucks in most of the north-flowing water from the Sound of Jura and waters to the south and, like a pinched hose, its narrows speed up the flow before spitting it out into the Atlantic to the west. As well as tidal flow and wind, the traveller must also negotiate the Hag, the whirlpool and standing waves for which the gulf is infamous. I read somewhere that a life-size human dummy in oilskins and buoyancy aid was thrown towards the whirlpool from a boat, whereupon it disappeared below the waves only to emerge hundreds of metres away with stones in its pockets! The suggestion is that it was dragged deep down and along the bottom, picking up debris on its way, before the violent current released its grip.

I have been through the Corryvreckan before and I have passed close by several times too, and if you get the planning right and are patient enough to wait for suitable weather, then a passage by kayak is entirely possible. Get it wrong, however, and the consequences can be stones in your pockets!

This time, though, I hug the mainland coast, well away from the Corryvreckan's clutches, then turn north-west and paddle over to the Isle of Shuna and on to Luing: two beautiful Inner Hebridean islands in the upper reaches of the sound, the former dark and thickly wooded, contrasting sharply with the grassy sward that is Luing to its west. In the calm water between the islands I share some sea stories with the skipper of the *Sea Dog*, who is taking fishermen to fish for skate, and I eventually put ashore at an exquisite spot at the top end of Luing itself. A grassy sward, flowers, the hardy russet brown Luing cattle mooching about, and views out in all directions over everything I love about the Highland seascape. Mountains to the east, islands to the west, glittering seas wherever is the sun, and a huge sky. I feel small here, insignificant, a tiny temporary orange dot on the ocean, a curiosity to some but invisible to most. As I move into view the sea creatures – the birds, the seals, the otters – lift their heads and watch intently then return to their lives as I pass. I am the quiet traveller, the slow traveller, the traveller passing through – and leaving, I hope, no trace. Giving very little, to be honest, and yet taking so much. Six days in, and my mind is starting to settle. The intensity of working life – the projects, the people, the deadlines, the expectations – is starting to fade. In place of that, a calmness I have not felt for a long time. It is in the solitude, I think, but also maybe in the moving, the journey. And not just any journey, but one under some *extremis*, with some challenge, some discomfort. And counter to that, there are these moments of deep pleasure as Nature folds herself around me. I am learning to accept a simpler way to be in the world, and I am loving the feeling of the open horizon ahead. A deadline of two months is no deadline at all really, not at Day 6 at any rate.

- Route: mouth of Loch Craignish, east side, through Dorus Mòr, then north to the north end of Luing
- Distance: 20 km
- Weather: rain, variable wind, cloud, some sun towards the end of day and showers, always showers
- Wind: Green/Amber. Force 2 to 3 gusting 4
- Sea state: bouncy, tidal, calm in the lee
- Events: crossing the Dorus Mòr
- Camping: on Luing.

Thursday 9 June, Day 7: Luing to Oban, 23 km. Wind: Green/Amber.

Quite a day. A day of contrasts. A day of tide, wind, messy seas, exposure, some fear, dog-tiredness, annoyance, delight and more.

An early start from Luing to catch the last of the tidal flood through the Cuan Narrows just to the north of my camp. The Cuan is another pinch at the top end of the Sound of Jura, and another tidal flow that you need to get right. The name 'Cuan' is Gaelic in origin, Scots and Irish, but has a varied etymology. It can relate either to 'ocean' or 'haven', and from my perspective on the water I can see how it might be both; travellers from the north, exposed as they are to the west winds, would see Cuan as a haven, while those, like me, travelling from the south, approach it as an opening out from the confines of the Sound of Jura, a new exposure to more open sea. This perspective has nothing to do with the etymology piece, of course, but I like to think that ancient names might have appeared from the seafaring perspectives and realities of the time.

The paper map in my deck case clearly dates back a lot of years, as I have written on it when and where Cuan's counter-flows and standing waves appear. I did that for my friend Robert's and my trip out here over 15 years ago. Nearing the narrows, I hear the sound of rushing water off my starboard side, and by luck I'm in just the right spot to shoot the gap. As I enter the channel itself, the land closes in on both sides and I can see that I'm flying along. The water beneath my keel is flowing north with all the weight of the Sound of Jura behind it, and it's a thrilling thing to release yourself to its power. A few houses dot the shore, but this early in the day there is no one about. It feels a bit lonely. Strange how a lived-in landscape with no people visible can seem lonelier than one uninhabited.

A breeze is building, and ahead is a line of breaking waves where wind is hitting tide. The line is around 100 metres deep, and fills the northern exit of the sound almost completely. This is a 'poke your nose out' moment; as I hug the eastern shore an occasional bit of relatively slack water appears and I'm able to slip out of the narrows unnoticed by the main race that would have gladly sucked me in had I been more foolhardy. These moments are all about respect, I think. The power of sea and wind are all-encompassing, and it's a foolish person who sees them only as something to conquer. You cannot conquer these things. With no power except that provided by your own body, you move through seascapes in a boat this size only with the permission of wind and tide. If you listen and watch for long enough, then hopefully you build enough wisdom so that you know when you can stick your nose out and when you should retreat.

Once I'm through the sound, a 3-kilometre paddle to the tiny island of Easdale lies in wait. By now, the forecast change in weather is in full swing. Blue skies have turned to grey and calm seas to a wind-blown mess, and a swell is building fast from the west. Swell and wind make for some interesting paddling, and when nature throws in a steep rocky windward shore the energy in the waves bounces back through the water to meet those incoming. At its best it creates a somewhat confused sea that needs careful concentration. At its worst, the same confusion can morph into what is known as *clapotis*,[1] which can churn the sea into a jumble of water moving in all directions with no discernible pattern. When it is like that, for me anyway, fearty as I am, it's a PLF (paddle like f..k) moment, and I just have to hold my nerve and my balance, and punch through to calmer water.

It's a wet paddle with wind on the beam up to Easdale. Here the small cluster of houses is also quiet, almost eerie, and I can see the old collapsed and long-flooded quarry from days past. Easdale – slate quarry *extraordinaire* back in the day, and now home to holidaymakers, a small resident community and the World Stone Skimming championships – slips by my port side, and a wide-open space looms ahead. Out of the shelter of the island, the wind-blown sea is grey with an occasional white breaker – lumpy and, to be honest, not welcoming. I perform a bit of soul-searching, feel my way forward into the lumpy stuff … and *breathe!* I have noticed that when I'm nervous I stop breathing. That's not good, eh? I then tense up, and my mind starts to play with my confidence. I don't think I'm a natural risk-taker, to be honest, not when it comes to the thrills of speed or jumping from heights, for instance, and my (very occasional)

1 A sea area consisting of the rebounds of waves from a seawall or cliff. The interactions of the water generate steeper and higher than usual standing waves. More about clapotis on 20 July.

darker dreams invariably involve some falling-type terror. When the sea rises, so does the fear; it's only natural, surely. The trick, then, is to find a way to overcome instinct with some rational thought, a bit of mindfulness if you like. My technique is to talk to myself, and when paddling solo this works a treat. I also now have Wilson, whose wee smiley face works to lift the mood. I do know that he is actually an it and is entirely inanimate, but what is imagination for if not conjuring up the impossible? My own voice becomes a third party, reassuring me that I've got this, that I have the skills to stay upright – and, most importantly I think, it reminds me to breathe. Yoga breath, steady, deep, settling, fuelling. To breathe is to calm and to calm is to untense, to stay flexible in the hips, to allow the boat to do what it wants to do, to ride the waves, to flex and tilt and surf and slap. And all I have to do is keep paddling.

The route ahead feels very exposed, and I'm alone off a rocky coast with cliffs and shores offering no get-outs. The wind is now on my back, the seas are growing in height, funnelling between mainland and the offshore island of Insh out to the west. The decision to go on is marginal, and I hold a line for a few minutes, feeling the water, testing my nerve and deciding what to do. In the end you have to commit or turn back. I decide to commit, and this then provides the most challenging hour of the journey so far. Wind behind and slightly to starboard, boat yawing about and needing regular bracing with the back of the paddle. Lots of self-talk: 'Relax, breathe, you've got this.' Then there is a moment, as the coast tilts a little more east, when the size of the waves begins to diminish – and relief and elation, a jumble of emotions, flush through me, and I feel invincible and, for a few moments at least, at home in the wind and the wet.

North Seil is lovely, a mass of small green islands covered in bracken and heather. I put in on a seaweed-covered shore for a break, check the mobile and give Leah a call to arrange our rendezvous. Oban, a small bustling Victorian town reliant now mostly on tourism, is around 13 kilometres away, and with the sea state as it is I think I might make it there today. Also, bigger winds are on their way later this afternoon and are due to last for a few days. A tropical storm way out west over the Atlantic is expending itself over western Scotland, so I need to get somewhere to give me options again. Any westerly shore will likely

Last of the calm weather in the Sound of Jura

Tucked away from the west on Luing

Campfire comfort

become too messy for paddling, so now, with Oban within reach, I can perhaps head out into the Sound of Mull and make headway there.

Leah starts to drive south from Inverness and I paddle on northwards towards the sound between the Isle of Kerrera and Oban town beyond. More messy sea to get into Kerrera Sound, then I pass a seaweed farm owned by a sister company of my old workplace, and finally cross some hidden point where the sea chooses to throttle back, and relatively calm water lies ahead. I put in at a slipway at Gallanach Beg, where the road comes close to the shore and where Leah and I have arranged to meet.

Over the years, I have paddled into hundreds of slipways and piers in the Highlands, but at this particular slip things take an unpleasant turn. A lady appears with a fierce face and an aggressive stance on my seeking access; this is something I have not experienced since before Scotland's Rights of Responsible Access bill was passed in 2003. She wants me, in no uncertain terms, to get off her slipway, and her sarcasm, open aggression and total lack of manners make for an unpleasant encounter. She claims the slipway is owned by the nearby recreational services company, and makes it clear they own the tiny rocky beaches round about too, and that I am not welcome. I try to converse politely and demonstrate that my intent is not to break any rules but simply to come ashore as I have done on slips across Scotland, and move my kit up to the road to meet my wife. But she has laid out her stall and seems unable to dial it back. Perhaps I should have held to the 'right of access' rules and/or put ashore at one of the tiny beaches close by, but in the end her aggression wins out. As I have no desire to have a great day ruined, I withdraw gracefully without expending any more emotional energy, make a note to follow up the incident with the Access Department at Highland Council, and then, phoning Leah to tell her of the change of plan, paddle on all the way into Oban itself. When I come ashore under the memorial at the north-west end of town, Leah is there to meet me and all is well.

- · Route: Luing to Cuan Sound to Easdale to west side of Seil to Oban
- · Distance: 23 km
- · Weather: cloud, grey and overcast, sunny spells later

- Wind: Green/Amber. Force 3 gusting 4, steady
- Sea state: everything from slight breeze to hard gusts take seas from calm to messy, and from Cuan onwards an underlying swell lifting everything up a notch
- Events: Cuan Sound, tricky seas to Easdale, and bigger still north of Seil. Meeting up with Leah
- Camping: B&B.

2 | Mother Nature rules

Days 8 to 16 – Oban to Eilean Giubhais

Friday 10 June, Day 8: Oban ferry to Craignure, on Mull, then north of Craignure, 2 km. Wind: Red.

Winds are up as expected today. Leah and I pore over maps, forecasts and ferries, and work out various options. There aren't many. The only way to paddle over the next few days is to get onto the lee side of mainland or island shores, as the winds are rising to gale force from tomorrow. Unseasonable stuff, it seems, and the sea out in the Firth of Lorne, between Oban and Mull, is white.

One option is to get a ferry to Castlebay on Barra and try to paddle up the east side of the Uists. That's quashed when CalMac cancel that same ferry for the next few days due to weather. Another option is to get a short ferry ride to the tiny port of Craignure on the south end of the Sound of Mull and haul my way north up the sound, hoping for a drop in the wind by the time I get to Ardnamurchan. CalMac has already put that ferry on 'warning' too, due to the weather, but today's afternoon sailing is on, so I go for it.

Leah drops me at the boat terminal and I use my kayak trolley to hump boat and kit onto the ferry. Out in the sound I see the wind and sea are far too fierce for a lone paddler, so it's clear I've made the right decision. Lovely moment with Leah waving from the point as the ferry passes through the narrows exiting Oban Bay. We're a couple of romantics.

So here I am, camped up in another idyllic spot just a couple of clicks north of Craignure. The rain has set in, and the wind is shaking

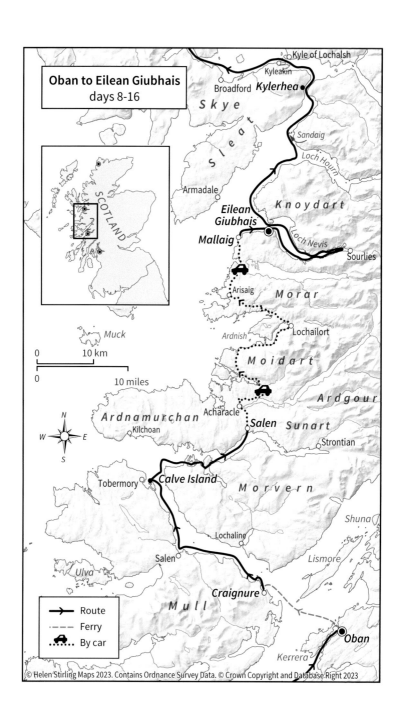

Oban to Eilean Giubhais
days 8-16

Kyle of Lochalsh
Kyleakin
Broadford
Kylerhea
S k y e
Sleat
Sandaig
Loch Hourn
Armadale
K n o y d a r t
Eilean Giubhais
Mallaig
Loch Nevis
Sourlies
Arisaig
M o r a r
Ardnish
Lochailort
M o i d a r t
A r d g o u r
Acharacle
Salen
S u n a r t
Kilchoan
A r d n a m u r c h a n
Strontian
Calve Island
M o r v e r n
Tobermory
Shuna
Lochaline
Salen
Lismore
Craignure
M u l l
Ulva
Oban
Kerrera

SCOTLAND

0   10 km
0   10 miles

N
W   E
S

Route
Ferry
By car

© Helen Stirling Maps 2023. Contains Ordnance Survey Data. © Crown Copyright and Database Right 2023

the tent with vigour. I pick moments between showers to get the tent up, and even manage a swim.

The journey continues. The weather has broken and looks challenging for days ahead. I am keen to keep moving, to maintain momentum, and it's likely the perfect settled weather I'd hoped for at this time of year is not coming any time soon. Instead, I'll need to take opportunities when they come, use the early mornings and late evenings when there is sometimes some relative calm, take some risks and, well, just keep poking the nose of my kayak into those marginal messy seas when they appear. I have dealt with the wind and the wet before – I live in the Highlands, after all, and am an ex-fish farmer to boot. Indeed, life at Kerracher in Assynt was very much an outside existence and was dominated by the weather. In addition, during Robert's and my Patagonian trip it had poured with rain every day and the winds had blown hard too. Back then, we had journeyed hopefully and had become resilient to what Mother Nature threw at us. So I remind myself that I have form and that I do know how to deal with adverse weather. It's a mind game, and resilience lies at the heart of it.

Why, then, journey? Why volunteer for being wet and uncomfortable and potentially dead? These are good questions, otherwise couched as 'What's the point?' The stock answer is 'Well, if you have to ask, you won't understand the answer.' But that response is a bit trite. Just because we add meaning to our personal actions, in this case a kayak journey, it doesn't mean they should become important or meaningful to others. We all journey after all, just in different ways.

So what is the meaning of journeying through the Highlands in this manner? I could have simply popped in a car and done the same. Except I couldn't. The nature of this journey is to expose myself to the sea, to the seaward, and therefore often the windward, sides of our islands. In so doing I had hoped that the act of moving, of journeying under my own steam with no distractions would release something in me to better allow proper observation of the surrounding environment and to think about some of the issues of the day as they relate to the Highlands. But it's interesting, too, to think about just how addicted we humans appear to be to the act of journeying, to moving on.

I sometimes think that moving on seems to be what we do, even what we *are*, perhaps. We all journey after all. Some do so close to home or out to safe organised places that feel a little bit like home. Others like to strike out further, and still others, the ultra-inquisitive few, disappear over horizons physical, emotional and intellectual. It is these horizons I find so addictive. Turning that corner, discovering the new, finding things out.

Maybe it's innate, and there is plenty of research to support that theory. Annie Murphy Paul, author of *The Extended Mind*, pulls together this and other neural research to show how our brains light up when we engage with moving through outdoor space and/or spatial reasoning. Much of our thinking, our learning and our creating is performed best, it seems, when the spatial reasoning parts of our brains are activated. One of the challenges of our modern intellectual ambition is that it requires elevated levels of abstract reasoning, and this is difficult for most of us – and certainly for me! However, by activating the hippocampus region of our brains (the part that, as well as accommodating our short-term memory, enables us to navigate through space by mapping out such abstractions into a spatial format), we become measurably more effective at capturing its meaning.

Moving through space, both physically and intellectually, lights up our ability to think, it seems.

And then there is the effect that moving around outside in nature has on us. Grasping book- or computer-based information, getting to grips with a complicated problem, seeking 'inspiration' are all optimised, too, when we go outside. Our brains come alive when we expose ourselves to natural, or at least natural-feeling, spaces. Why? Paul suggests that we are 'tuned in to the natural frequencies of the organic world'. This is a result of the many thousands of years of our evolutionary development when such tuning in could be the difference between life and death. Understanding the weather, spotting edible plants, avoiding the sabre-toothed tiger.

Professor Clive Gamble explores our human wanderings as 'timewalkers', and uses an extensive archaeology and palaeontology to paint a picture of early hominids and modern humans walking and boating out from their ancestral homelands towards pastures new. We humans occupy every ecological niche because of such journeying. Our culture, language, tools, cooperative abilities, familial ties and cognitive sophistication all allow us to experiment, to take risks and to move outside the comfort of our known worlds. At the most extreme, humans have journeyed even out to the ice-covered islands of the high Arctic and made them their home. In the Pacific, the canoe peoples ventured out beyond the ocean's horizon to discover Hawaii, Easter Island, New Zealand and more. We have the journeying gene, it seems, as part of our evolutionary baggage.

Likely driven initially, back in our distant past, by a need to discover resources just to stay alive, we do it still. Workers move to where the work is, across countries, across continents. Economic and political refugees move because they must, and we settled wealthy folk in the west move when we wish to and because we can. It is a restless experience, this desire to move, what the American novelist Kurt Vonnegut calls (quoting Bokonon) 'dancing lessons from God'. This urge to move mirrors a constant desire for change, for something new, something better perhaps. We move, we journey, because that is what we are. Timewalkers indeed.

On a personal level, I have always felt a strong need to peer over the horizon, be it physical or intellectual, and while living with such restlessness cannot be easy, it does mean that interesting and often unexpected things happen. At a micro-level, on a kayak journey for instance, wonders appear when you stick your nose out. It would be easier and safer to shelter behind the headland, but when you take the risk, when you move into that marginal space between comfort and terror, a new world opens up. The result, in this micro-case, is a journey both through space and into ideas: a purposeful wander through an island nation stuffed full of wonders, where ecological, social and political issues abound, and where the physical journey strips away one's comfort zone and challenges one to think and live a little differently. It's only Day 8, but my mind is settling to a new rhythm and I have many horizons ahead.

- Route: Oban to Craignure (ferry) then 2 km north of Craignure to stony beach
- Distance: 2 km
- Weather: wind and wet
- Wind: Red. Rising to Force 7 gusts from the west, and rising again tomorrow

- · Sea state: not for me!
- · Events: dodging the torrential rain showers; swimming anyway; first real lockdown in the tent
- · Camping: southern end of Scallastle Bay.

Saturday 11 June, Day 9: Weatherbound north of Craignure, 0 km. Wind: Amber/Red.

Wake up feeling lucky. Lucky that the tent did not collapse under the onslaught of the wind gusts last night, and lucky that I have a lovely wife who is prepared to drive out to meet me on occasion if the weather stops me in my tracks. Last night it rained for hours. Not just rain, but torrential stuff, every bit as wet as Patagonia all those years ago. And the wind! I could hear the roar as each new gust careered down off the hills looming to my west, thrashed through the trees and shook the tent like a rag doll. I lay awake for much of the night watching the poles bending hard, hoping they would not break and making plans for if they did.

With the daylight, I poke my head out of the tent flap between downpours and am met with a sea white with foam and a westerly curling round the hills and thumping south down the sound. No way of paddling today, so am hunkered down, reading, writing and planning. The forecast is for these strong westerlies to last for several days. This makes west coast paddling untouchable, a death wish really, and I quite like being alive. There is a small window of lighter wind in the very early mornings over the next couple of days, so perhaps I need to take advantage of that. It is light by 4am, so maybe that's my way forward. I could work my way up the Sound of Mull then snatch a weather window to get over and into the long reach of Loch Sunart. It's a dead end there, of course, and I would need Leah's help to move on from there – but at least I'd be moving now, not sitting huddled in a tent.

So that's the plan: to head up towards Tobermory tomorrow, 25 kilometres away, suss the conditions when I'm on the water and make marginal decisions as I go. Nothing changes, it seems, except everything all the time …

Managed to get the tent moved in between showers so I now have more protection from the winds. I have eaten, read, yoga'd, fixed my waterproof trousers, eaten some more, planned and constantly peeked out the tent hopeful for less wind and wet. The rain, however, has been incessant. It's not cold, though, so that's a blessing.

- Route: stationary, pinned by weather at camp spot north of Craignure
- Distance: 0 km
- Weather: windy and wet
- Wind: Amber/Red. Force 5 to 7 all day and incessant rain
- Sea state: white!
- Events: the weather
- Camping: southern end of Scallastle Bay again.

Sunday 12 June, Day 10: Craignure to Calve Isle (east of Tobermory), 25 km. Wind: Amber.

Early start; up at 4am, on the sea an hour later, wind in the face, rain and more rain. Trying not to rush everything or get too anxious. 'Soaked in seconds' type rain is only minutes away all day. When it comes, the sea is pockmarked with large drops, my shoulders and arms are getting wet inside my jacket, and it's head down and paddle on. It's warmer in the boat, under the spraydeck, than standing about on the shore.

Work hard today. Wind and tide on the nose. No choice really, as the only way forward is up the Sound of Mull, and with the weather windows as they are I have to prioritise wind over tide. It's a spring tide, maximum flow, so I only make around 4 kilometres an hour up to Fishnish. A wet trip with constant spray as I push the boat against the incoming waves. Grind my way north up to Salen[2] on Mull, where the sun comes out at last. The relief of sun after the rain is a joy. It's only 8am and I've already paddled for three hours today. I need a reward, so I haul the boat up over the weed and tramp up to the lovely old Salen Hotel for breakfast and to dry out.

---

2 In my journeyings I visit more than one Salen, which, once you know that the name probably derives from the Gaelic word meaning 'little arm of the sea', is unsurprising.

Stormbound in the Sound of Mull

Whitecaps and incessant wind in the Sound of Mull

Wilson still smiling

An hour later and full of fried food, I paddle on towards Tobermory with new energy. A beautiful day forms, still blowy, but I'm mostly in the lee now so less of the struggle of the early morning, and as a bonus I'm paddling through a birding bonanza: nervous Canada geese with goslings, Arctic terns squabbling over small islets, regal mute swans, goosanders, bubbling calls of curlew, piping oystercatchers and sandpipers, stately herons, redshank, gulls, greylags, mallards, and lots and lots of eiders. The geese are still growing their new feathers, so some of the adults cannot fly at present. As I pass, the parents attempt to draw me away with a clumsy flapping across the water, enticing me from their goslings and potentially sacrificing themselves in the process. They don't know I mean no harm, of course. If the birds, seals, and otters see me enough in advance they don't seem to panic, but if I appear suddenly round a corner then that sets them off. I try to move mindfully through the islets, especially at this time of year when parents are caring for children, but I can't get it right all the time.

I am camped on Calve Island off Tobermory. I decide not to go into the village, as I have been there many times before and there's rain coming (again), so I prioritise getting the tent pitched in the dry. I fail; the rain has come hammering in as soon as I hit the shore. I sit hunched like the animals, back to the wet, for 20 minutes, waiting for the squall to pass. Water running over my head and shoulders and hands, not too cold though, and observing the patterns of drops on the water and tide rising and trying to be patient.

A good day though, plenty of mileage under the keel, and while I'm physically tired and have been wet for much of the day, the rain and the sunny spells have brought the vivid greens and blues of the islands to life. I am well placed now, as planned, to get over the sound in tomorrow's early-morning weather window. Creeping forward.

- Route: Craignure to Calve Island off Tobermory
- Distance: 25 km
- Weather: wet, biblical downpours and sunny spells
- Wind: Amber. Force 4 gusting 6, but more sheltered shore available after Salen
- Sea state: wind-driven waves, small but hard work; tide against me, too, all the way to Salen

- Events: the paddle itself, torrential rain, sun coming out, early start, many birds, breakfast at Salen, camping spot on Calve, view out to Ardnamurchan and entrance to Loch Sunart
- Camping: on a bog.

Monday 13 June, Day 11: Calve Island to Salen (on the mainland, by Loch Sunart), 20 km. Wind: Amber.

Alarm bleeps at 3.30am but I'm already awake, fretting about the crossing ahead: no crossing, no journey. The pressure, self-induced, not to turn back means my only option now is to get across the sound and into Loch Sunart. The idea of turning left and attempting a round of Ardnamurchan Point is a death wish for the next few days, and especially so as I'm on my own. It's wild out there. Turning right into Sunart will eventually give me shelter from the wind, but not the incessant rain.

Unlike yesterday's battle against wind and tide, today's wind should be on my back once I have turned east into the mouth of Sunart. The tide, too, is just right. Slack for the crossing, high for getting the boat into the water from the rocky shore, and still rising by the time I should be getting into Sunart.

At 3.30am it's not really light, but by 4am it is, and I get busy packing the wet tent and all kit into the boat. Wind, as forecast, is westerly and looks a Force 3 out on the water. Calm on the Calve Isle side, but white breakers on the eastern shore where I'm headed. It's only a 2-kilometre crossing, but at 4am on a dreich[3] day, on my own, with rain, low cloud and mist and a windward shore ahead, I'm feeling somewhat exposed.

Then, packing done, I'm on the sea. Everything changes then, from solid land to moving water as I pull rapidly out from shelter into the sound; no get-outs, committed. First half of the crossing is fine, and then the wind and waves rise and a confused bit of sea surrounds the boat. I talk to myself and that helps calm the nerves. Remembering to breathe, that's the key, relax the hips, let the boat do its thing, focus on

---

3 ... which means precisely what it sounds like.

the waves coming at me from the beam, read the gusts … all that. At the point, the westerly swell has risen beneath the wind-driven waves and is meeting a bit of tide still rising north up the sound. Then the turn, wind on the back, an unsteadying sea picking up my stern and surging me east. While I focus on staying upright, the world has disappeared. Thick clag leaves Mull invisible – and the north side of Loch Sunart, only 3 kilometres away, is gone too. Rain starts to pelt down again.

The coastline from Auliston Point is steep, craggy, tree-covered and pouring with wet. Focused on staying upright, eyes on the waves, I can only glance up on occasion, but can see what a spectacular place this is. New streams falling through the dark tree-covered cliffs, and everywhere the rush of wind and wet. I feel a pang of loneliness at one point as the rain and mist, looming landscape and tricky sea seem to be conspiring against me. That's a piece of nonsense, of course, as Mother Nature doesn't care whether I'm here or not, but perhaps the very fact that I habitually refer to nature as 'mother' means I'm hoping she does. I make a mental note not to volunteer to lie on any psychologist's couch any time soon.

Four kilometres of balancing act, and I enter at last the sheltered inner loch of Droma Buidhe alongside the Isle of Oronsay. The sea quietens and calm ensues, but the rain keeps on. The underarm and shoulder sections of my paddling jacket are well and truly wet inside. Not good. Still six weeks to go, and already my jacket is not waterproof. I need to do something about that soon.

In the bay there are a couple of yachts sitting at anchor hunkered under the hill in the lee of the westerly; the rain is pouring off the hills and the green woodlands clinging to the slopes are threaded white with streams. A long old wet paddle follows, aiming for the village of Salen, and hopefully to meet up with Leah at some point for a lift to Mallaig. As my hopes for rounding Ardnamurchan this trip have been dashed by the forecast I must pivot, and work a solution that enables me to keep the journey going. Jumping forward to Mallaig is a logical move, I think. But the coast between Sunart and Mallaig, all westerly facing, is awash with weather. From Mallaig I can head into the shelter of Loch Nevis, to see out this period of wind and wet, then go back out to sea again, to continue the journey north towards Skye and beyond. So for now, Salen on Loch Sunart it is.

It's a tired paddler who sneaks in under the hill at the village and, still in the rain, lifts boat and kit up above the tide. But then the wee store on the pier opens and the Kiwi couple running it could not be more welcoming. Coffee, toasties, recharged phone, and I call Leah, who sets off from Inverness. It's only 9am, after all, and I feel as if I've already had a full day while everyone else in the world is just getting up. In time, the rain stops and I walk up the road to Acharacle at the head of Loch Shiel to stretch my limbs and to meet Leah coming the other way. We meet, collect boat and kit, and head off on the hour's drive to Mallaig and the luxury of a B&B.

Mallaig is *wet*! Our room becomes a drying room with washed wet kit dripping onto towels, but to be out of the rain for a short while is great.

- Route: Calve Isle to Salen on Loch Sunart, then relocate to Mallaig
- Distance: 20 km
- Weather: wind and wet, clag
- Wind: Amber. Force 3 to 5 and rising to 6
- Sea state: lumpy and wind-blown out in the Sound of Mull, especially rounding Auliston Point, then windy but okay once past Isle Oronsay.
- Events: early start, entrance to Sunart, wet
- Camping: B&B in Mallaig.

Tuesday 14 June, Day 12: Mallaig to Sourlies Bothy, 23 km. Wind: Green/Amber.

Today has dawned wet and windy as forecast, and after baulking at the idea of getting going in this weather, I bite the bullet, pack the boat in the pouring rain and paddle off for Loch Nevis. Leah stands out in the dreich, waving me off, and I'm alone again.

A long 20-kilometre paddle into Loch Nevis and all the way up to Sourlies Bothy at the head of the loch. Tide timed perfectly, though – just a bit of luck really – so I'll avoid the long drag of boat and kit up to the bothy itself. Great tidal flush through the kyle narrows with plenty

Salmon farms – the blue food economy at work in the Sound of Mull

Building resilience to the incessant wet

Tomorrow's crossing into Sunart from Calve island

of swirls and boils for a few hundred metres into the inner loch and then, unexpectedly, calm.

Sourlies is a wee haven of relative dry from the incessant wind and weather of the past few days. I say relative dry, because the air is so full of moisture and the bothy such a simple stone-built structure that 'dry' actually means 'damp'. I think I need to develop a lexicon around the varying levels of wetness experienced when sea kayaking. 'Dry' is when you're back at home and not expecting to be wet any time soon. 'Damp' is that all-pervading clammy wetness on your clothing and kit from the moist air and high humidity. 'Salt wet' is when salt-laden clothes become wetter overnight as the salt soaks up the moisture in the air. 'Wet' is rainwater on the head, legs and hands. 'Wet-blast' is the wind-driven rain that fires through zips and gaps in your clothing. And finally, 'soaked' is, well, when you go swimming or capsize.

It is rather wonderful, though, how hardened you become to being wet all the time. It is either that or run greetin' for home, but I find that after a week or so of exposure I stop noticing it quite so much. I go bare-footed most of the time, so my feet are always wet and I'm used to that. Hands, too, are always wet in this kayaking and camping game, and once you realise that your skin is waterproof, there is a peace of mind that comes from accepting a new normal. On the kayaking trip in Patagonia in 2005 I went fishing in a glacial-melt river gushing down from the mountains. With forest hemming in on each side, I stood in the middle of the stream of icy water to get the space for the fly rod. As I moved out into the river a startling new vista appeared; a couple of kilometres upstream, a vast deep blue glacial snout thrust out from the cliffs, the evergreen beech forest, unchanged since the close of the last ice age, softening its edges and the river, cloudy with glacial silt flushing whites and greys through it all. It was pouring with rain, and almost immersed in water as I was I had a rather wonderful vision of my place in this wilderness: a tiny dot, barely visible and certainly irrelevant in the grand scheme of nature around me, but comfortable in that place, in that wet. ...

Now, in the Sourlies Bothy, Dan appears. Dan is from Brighton and he's humping a huge pack over the Cape Wrath Trail. He seems a bit on edge from the incessant rain and the river crossings, and he has no cooker, therefore no hot food. I make him tea and a meal, and in return he tells me all about the Brighton Bohemian scene, which just makes me wish I was 20 years younger.

- · Route: Mallaig to Sourlies Bothy in Loch Nevis
- · Distance: 23 km
- · Weather: wind and wet, very wet
- · Wind: Green/Amber. Force 3 gusting 4
- · Sea state: a bit of everything. Some light lump out of Mallaig and down the channel open to the west; flat calm in the inner loch
- · Events: Leah waving goodbye in the rain, the drenching mid-loch, bothy haven, bohemian Brighton
- · Camping: Sourlies Bothy.

Wednesday 15 June, Day 13: Weatherbound at Sourlies, 0 km. Wind: Red.

Stormy weather has struck the coast as expected, so for me today is a rest day. Dan has left early so I dry my kit, clean the bothy, do yoga, read, write, eat, and gaze out at my surroundings. This really is a stunning place to sit out the turbulent seas out west. Tucked deep into the mountains at the head of Loch Nevis, the bothy is almost invisible until you're right on top of it, dwarfed as it is by the precipitous slopes of the surrounding mountains. Rearing above me is one of my favourite Highland hills, Sgùrr na Cìche. From the sea it's a perfect conical peak, and the westerly ridge is the perfect approach. From the top, so much of the Highlands is visible, and as you gaze west, the jagged ridges of the Skye Cuillin and Rum are full of the promise of adventures ahead.

Today, though, the cloud is low at just a couple of hundred metres, and the mountain tops have disappeared into the clag. It is also wet, very wet, for most of the day.

Late afternoon, and two young lads from Poland arrive. They too are wet from their rain-soaked tramp that day, but unbowed and obviously delighted to find a semi-dry bothy for shelter. Fluent in

English, sharply intelligent and brimming with the possibilities of youth, they are in the last couple of days of their Cape Wrath Trail adventure for this year. Professionals working in IT and finance and operating from Warsaw and London, both are also deeply involved with the Ukrainian refugee crises, the conversation around which lasts long into the evening. Two more inspiring and intelligent companions I could not wish for.

We enjoy the relative calm of being tucked away in the inner loch all day, and a whole half-hour of dry with a light breeze to help dry out some kit before the rain settles back in again. Chat with the Polish lads till late. Once they discover I've been married for 35 years they want to know the secrets of how to find and hold onto the right life partner; endlessly swiping right is leaving them with too much choice and not enough commitment, it seems. I find it hard to think of two more eligible bachelors, but clearly they are struggling to find a match. I tell them I have no idea, really, as I'd started dating before mobile phones and global choice of partners at the press of a button were a thing. But they want tips anyway, so I give them the following to chew on: share values and respect differences, give space, develop a shared vision on how you want to live life, share adventures, expect challenges and work through them, commit, plan ahead, don't get fat, accept that some compromise is normal, accept that the connection between you changes over time, and don't look for the lightning bolt at the start … it may be someone you already know. I have more to give (poor them!) but they seem happy with that as a starter for ten. Good luck, lads!

Repaired the boat today, with Gorilla tape, no less.

- · Route: weatherbound at Sourlies
- · Distance: 0 km
- · Weather: wet, wet, wet
- · Wind: Red. Force 6+, and even windier outside the loch
- · Events: the Polish arrival, saturated air so nothing drying
- · Camping: another night in the bothy

Thursday 16 June, Day 14: Weatherbound at Sourlies, 0 km. Wind: Red.

Wake at 5am and, without disturbing the Poles, enjoy a quiet yoga session on the raised platform that's my bed, then an all-over body wash in the stream in the rain. The lads wake later and we all three go swimming, the guys squealing at the cold like young boys, but doing it all the same.

My new Polish friends leave for Inverie, a tiny off-road community on the Knoydart Peninsula, and their journey's end. I strike up the hills to climb Sgùrr na Cìche, a Munro.[4] The wind and the wet mean the day remains unkayakable, so the hills seem a good compromise. The cloud base is a touch higher than yesterday so I enjoy some dry spells on the way down, but the top section of the climb is especially wet. No view, but feels good to be moving, and wonderful views down Loch Nevis once back out of the dreich. The Rough Bounds of Knoydart and the hills south to Arkaig are all grey, clouded-scudded, dramatic and devoid of any settled human life. A big empty space in the heart of the Highlands.

Back at the bothy, two more walkers arrive. Alastair moves on, but James stays; he's a molecular biologist working on covid vaccines. I've been lucky with interesting companions here at Sourlies.

The rain continues and the hillsides opposite and behind are running white with new cascades – and a stream at the back of the bothy has started to flow inside it! The day is humid, the grass underfoot also feels unusually warm, and the air is full of moisture. All my dry kit gets wet again just from being exposed to the air. This is turning into one wet and windy trip. My eyes are constantly aloft checking for the next rainstorm, and when it pours it really pours.

- · Route: weatherbound at Sourlies
- · Distance: 0 km
- · Weather: wet and windy. Very strong on top of the Munro
- · Wind: Red. Force 6+
- · Events: climbing Sgùrr na Cìche, endless rain
- · Camping: another bothy night, rain and wind incessant.

---

4 A Scottish mountain of altitude at least 3,000 ft (~914 m)

Wet exit from Mallaig

Downpour and the jacket starts leaking, Loch Nevis

Friday 17 June, Day 15: Sourlies to west of Ardintigh Bay, 12 km. Wind: Amber/Red.

Today the sun has appeared for the first time in days. Like being born again. The air is still very humid and rain is coming and going, so getting kit dry is still a challenge. But the sun is back!

James has left for Inverie so I'm alone again and keen to get on, but the tide and wind are all wrong for me to paddle out of the inner loch. In addition, the Force 5 gusting 7 from the west all night have battered my tent, where I was storing my kit, and I now have two broken poles. Luckily the jagged edges have not broken through the flysheet, so I use the day to repair those and then dig a trench in front of the bothy to allow the water to drain past rather than in. Good honest graft.

James reappears, having been spooked by too much water at the river crossings; his video shows a raging torrent, impassable unless you wanted a proper soaking.

At 8pm the high tide and a slight drop in the wind give me the window I want.

The next two hours on the water, now titled 'Escape from Sourlies' are a bit of an adventure. No swell, just wind-blown water, but against a Force 5 gusting 6 on the nose this is a 'grind it out' piece of paddling. I want to get to the narrows on the slack of high tide. If I'm late the outgoing flow will slap up against the incoming wind and the sea will be unpleasant. The light is fading, wind in my face, waves washing water over my boat and rising as I near the shallow ground at the kyles. Grinding it out.

Through the pinch and level with the tiny cleft in the shoreline at Tarbert, the fetch shortens and the wind and sea begin to diminish. Still gusting, but a much drier ride – and, unbelievably, the rain stops! Paddling past Tom 'Moby' McClean's place at Ardintigh, I see the sperm whale-shaped boat, infamous throughout the Highlands, on the beach. Moby, a veteran soldier and adventurer, has been running adventure courses out of Ardintigh since the 1970s, and he is perhaps best known for his extraordinarily daring venture to claim Rockall, a lonely rock pinnacle way out in the Atlantic, for the UK in 1985. He has rowed and sailed across the Atlantic several times, and remains one of the UK's most colourful adventurers. *Moby* the boat is an architectural wonder, once intended to wallow across the Atlantic but now slowly decaying on the beach. McClean's life adventures make mine pale in comparison, and I love that we have such characters close to home. I paddle past the somewhat dishevelled but still smiling whale and park up on an entirely unexpected perfect pebble beach and camp spot at 10pm. Very tired but pleased to have broken out of the loch and be on my way again.

One of the Loch Nevis salmon farms is just offshore, and probably because of their investment in building masts in remote areas I have a mobile signal for the first time in three days. Great to get a forecast – but not so great that the weather is to remain challenging for days yet, it seems. This is supposed to be June, full of sun-kissed days and calm seas. Instead, it's unusually windy, wet and rough out here, and I can't help wondering if this is climate change coming home to roost.

- Route: Sourlies to west of Ardintigh Bay
- Distance: 12 km
- Weather: dry … mostly
- Wind: Amber/Red. Force 5 gusting 6 on the nose
- Sea state: lumpy at all headlands, however minor; confused at the narrows, then calming under the hills west of Tarbert
- Events: the sun appearing, late evening paddle in the mid-June gloom
- Camping: perfect spot for a tired paddler.

Saturday 18 June, Day 16: Ardintigh Bay to mouth of Loch Nevis, 8 km. Wind: Amber/Red

Another day dominated by difficult weather, mostly wind but also some wet. My neck is giving me gip today, but my shoulder at least is holding out.

Paddle in a mix of sunny spells and downpours up the west side of Loch Nevis, seeking out shelter from the wind still blowing hard from the west. Outgoing tide assists a little but still some wind on the bow and occasionally rushing down off the hills through any gaps to the west. Take a break at the turn and can see the sea ahead is surging and white in the westerly 5 to 7 as forecast.

I'm aiming to nose out slowly into the weathered sea, creep around the corner and get a feel for the conditions while still being able to retreat if need be. So windy! Heart in my mouth, I nose the kayak out into a very different world. The fetch is long from Sleat, several miles away, and the wind has been blowing hard now for over 24 hours. Out of the lee of the land and into waves breaking, sloshing up the rocks and back, gusts whirling in, boat lurching and slapping, and suddenly I'm feeling very exposed again. A hundred metres on and the swell appears, lifting the seas another notch. My self-talk gets me round the first point to a tiny corner where I can scramble ashore, drag the boat up the rocks and take a walk over the rough coastline to look ahead. Gazing out from 100 metres up off the shore, every now and then the conditions look doable and my heart lifts – but then it all turns white again. Sun and rain coming and going. I have spotted a small island, Eilean Giubhais; behind it there's an even smaller bay, which looks to have enough shelter for me to get ashore. I decide to go for that.

Scared the proverbial out of myself in doing it, but got behind the island and put in for the day. I'm not going any further in this sea – it's a death wish.

Manufacture a campsite on the wet lumpy ground using cut bracken to create a flat enough base, and once more feel shades of the Chilean journey, when spending a couple of hours cutting into the dense undergrowth just to make space for our tent was a daily task. Feeling dog-tired now, mostly due to the battle against the wind, I get the tent up just as the rain sets in. It feels endless, this always chasing windows in the weather. But finally I crawl into shelter, and that's me now for the day. It's only 3pm, and I feel like I need to sleep for a week.

The wind is going round to the north-west tomorrow, still strong, but there is a small window of lighter winds for a couple of hours very early morning which I might use to get a bit further north. Mother Nature's in charge, and I move when she lets me.

- **Route: Ardintigh Bay to Eilean Giubhais**
- **Distance: 8 km**

Escaping the gales at Sourlies bothy, Loch Nevis

Inner Loch Nevis from Sgurr na Ciche

It never rains…………..

Camping in the rough bounds

Slow escape from Loch Nevis

- Weather: sun, cloud, rain, wind
- Wind: Amber/Red. Force 5 to 7 ... relentless
- Sea state: okay in the lee of Loch Nevis, then nasty
- Events: taking on the sea to Eilean Giubhais
- Camping: challenging, on a slope, wet underfoot, but splendid view of the wind-blown sea.

3 | The art of wonder

Days 17 to 22 – Eilean Giubhais to Luskentyre

Sunday 19 June, Day 17: Mouth of Loch Nevis to south of Doune, 8 km. Wind: Green/Amber.

Another short day that doesn't feel short at all due to the challenges within. I take advantage of the small weather window early morning, up and on the water by 4.30am and manage to get clear of Loch Nevis at last. I'm getting better at this stuff.

Paddle across what yesterday was an untouchable wind-strewn sea. No sun to see this early, but broken cloud and an early light in the east is all pale oranges and yellows and blues. Kayaking alone while the rest of the world sleeps is a special time to be out. Once I am round the islands off Sandaig Bay I find that yesterday's wind is still in the water, and I take on some lumpy seas. I keep my nerves in check and put ashore at the tiny hamlet of Doune, an off-road community on the west side of Knoydart. When I land on the cold north-facing shore at 6.30am, the wind is building from the north, there is no one about and I start to shiver. This is not the place to settle with a strong northerly blow coming, so I steel myself to come back out into the lumpy sea and paddle back south to the bay I've seen on my way in. Low tide, of course, and the most boulder-strewn and slippery shoreline yet, so a long old haul of boat and kit is the result. Camp inside an abandoned shieling which provides some much-needed shelter from the rising north-westerly. Two of the new tent poles have broken already.

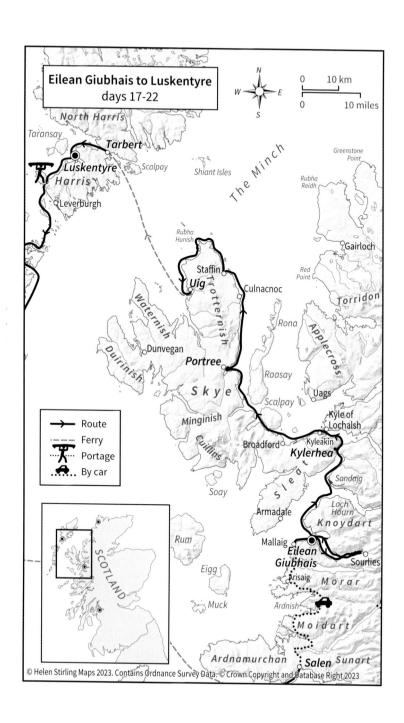

Eilean Giubhais to Luskentyre
days 17-22

N
W ← → E
S

0 — 10 km
0 — 10 miles

North Harris

Taransay

Tarbert

Luskentyre

Harris

Scalpay

Shiant Isles

The Minch

Greenstone Point

Rubha Reidh

Leverburgh

Rubha Hunish

Staffin

Uig

Culnacnoc

Gairloch

Red Point

Torridon

Trotternish

Waternish

Rona

Applecross

Duirinish

Dunvegan

Portree

Raasay

Skye

Scalpay

Uags

Kyle of Lochalsh

Minginish

Cuillins

Broadford

Kyleakin

Kylerhea

Soay

Sleat

Sandaig

Loch Hourn

Knoydart

Armadale

Route
Ferry
Portage
By car

SCOTLAND

Rum

Mallaig

Eilean Giubhais

Sourlies

Eigg

Arisaig

Morar

Muck

Ardnish

Moidart

Salen Sunart

Ardnamurchan

© Helen Stirling Maps 2023. Contains Ordnance Survey Data. © Crown Copyright and Database Right 2023

Dodging about alone in this difficult weather and on these messy seas and humping stuff up and down rocky shores, today I realise I'm actually settling to this way of being. I am starting to relax. Perhaps it is the Henry David Thoreau I am rereading, but thoughts are turning to the benefits of living simply, of doing without too much stuff, and to the restful power of being alone for a while, of solitude.

This journey does need stuff, of course, and some of it is fairly sophisticated: tents, kayak, stove, and of course the phone. However, I don't have many clothes, and so long as I have a good book and a notebook for writing I really don't need that much; I wash body and clothes in streams, eat simple foods, sleep, journey, repeat. The simplicity of it all is releasing, and the drudgery of endless washing-machine loads and hedge trimming-type tasks at home is gone. I've had very mixed weather, so it's been a challenge to dry clothes, but you take the opportunities when they come, and put your wet clothes on your body to dry when they don't.

As I write, the sun has appeared, instant warmth through the tent, even though it is only 10 degrees out there, and soon I will be laying my kit out on the shieling walls to dry off at last. My tent is small but not too small, and is perfect for this trip with a little more space for storage than any tiny one-man mountain tent. Everything I need is under cover, and I can even do my yoga too, albeit not the standing poses.

I love the simplicity of it all. As a baseline there is the journey – open-ended as it is by destination and more or less by time – and this structures my days and provides purpose. Everything after that is 'mission', and describes what I hope to achieve during the journey. This includes a range of activities such as: engaging with the sea and with wildlife, climbing the hills, noticing small things, looking up at night, moving with the wind, being brave when required, and making decisions thoughtfully. And I also want to listen to my gut-feel when danger looms; not to rush, but to enjoy stillness, take time to wonder, to observe, to create memories, to enjoy and experiment with the solitude, and be thankful for the life choices and chances that have enabled this journey to happen. My strategy is simple: to keep moving but not be afraid to turn back, to use my kit and outdoor knowledge well, to keep fit and flexible, to eat sensibly, and to move as the sea and the wind permit.

When you start thinking and writing about it, there is quite a lot going on, quite a lot to think about. But, Thoreau-like, much of it is quiet, out of the way, often personal, invisible to most – and impacting, I hope, on no one. Leaving no trace.

But what about the solitude? Am I enjoying being alone? I am discovering that solitude is not loneliness. I don't feel lonely at all. You could argue that with mobile phones always available, I am never really alone. But my policy is not to engage with the mobile as I would in normal life. I have determined to leave it switched off most of the time, and use it only for forecasts and essential texts. Some of my friends have Whatsapp'd me, which is lovely; they're showing an interest. I debated that one and decided it would be a bit mean not to acknowledge their contact just because I'm seeking solitude. I am blessed to have friends who care, and as some of them are paddlers themselves they understand the exposure of kayaking out here on your own. So I reply, and it turns out it is a lovely thing to have that contact. I do not rely on it for emotional support and most of the time the phone is indeed switched off and tucked away deep in a dry bag. It is comforting, though, to know there are folk out there thinking of me.

But the phone is useless when on the sea, and unusable in remote glens and shores where there is no signal. I am of a vintage where mobile phones did not exist when I began to roam Scotland's vast open spaces. We learned self-reliance and compass work, and we read the weather by looking at the sky, and read the tide by watching the shore. When things went awry we had to fall back on our initiative and deal with it there and then. No Help button or personal locator beacon, no what3words app. When you were out there, you were out there.

And now, bouncing about in the waves, fear rising and with 'what ifs' careering about in my head, I feel exposed and often fearful, yes – but not lonely. When remote camping too, mostly far from other people, I am alone, tucked away in my tent on some faraway rocky shore. I'm alone there, too, but again not lonely. Perhaps being blessed with family and friends means I know I'm not alone in the grander scheme of things, and I expect to return to real life in six weeks' time. But until then, it is solitude that is the open and close of each day.

Tucked away on west Knoydart

Dawn on the Knoydart edge, heading north

Early weather window across Loch Hourn

Being alone, especially for an extended period and especially too perhaps when travelling within such an intense natural context as the sea, allows me time to notice the world around me differently. When I'm with friends, my trips into the mountains or onto the sea are full of chat. Chat about our lives, politics, work and family goings-on. The mind is partly present in the place through which we are travelling, but invariably it is focused mainly on the chat, on the social interaction. In contrast, out here alone on the water, in the camp, on the hill, I find I can focus entirely on the present at last. Some folks do their thinking while out running, walking, or biking. I don't. My thinking time comes later, when writing, or often late at night or early morning. Outside in nature, I find myself utterly present and in the moment. There is, after all, so much 'thinking' that the natural world offers. Within touching distance there is geology, ecology, biology, evolution, social change, botany, migration, oceanography, land use, climate change, meteorology, politics. It is all there if one takes the time to look. Thomas Huxley, biologist and Darwin's champion – his 'bulldog' – sums it up well: 'To a person uninstructed in natural history, his country or seaside stroll is a walk through a gallery filled with wonderful works of art, nine-tenths of which have their faces turned to the wall.'

I wander this gallery, but hopefully do not do so empty-headed.

There is wonder, too, and wonder is something sadly lacking in our Google/Wikipedia-driven world. It is possible, allegedly, to know everything at the touch of a button, or at least to believe we do. But we need to leave room for the wondering too. John Muir, Scot and father of national parks, treads that line beautifully in his writings. He grew up in the mid-19th century, which saw religion and science competing for space, and his intellectual blossoming merges the two in writing that I find irresistible. He is a both/and type of guy, I think. God creates the world but then natural processes twist and shake and re-form and make it what it is. He is not afraid of expressing wonder in his writing. Wonder at the sheer magnificence of Yosemite Valley, at original creation, at the glacial pace of change, at the power of wind and water to shape the ground upon which we tread. I love him for that honesty, and in these days of knowing it all, a little room for wonder – dare I say reverence – of the as yet unknown, the forces of creation, the creatures still to be discovered, the environment as we once knew it, the world as it could be, might be a good thing. Perhaps a healthy dose of wonder might influence a different groupthink, a different politic around today's environmental agenda and a different set of actions as a result? Goodness knows we need it.

Does any of this sort of internal musing matter? Does the power of simple living and the embrace of solitude mean anything really? Well, to others perhaps not, but to me it matters because it helps me guide the way I choose to live, the choices I make and the work I undertake. There are contradictions within all that of course; life is complex and requires interaction with others, and sometimes you must compromise a little just to engage. But at heart it is our value systems that do most to guide our hand, determine our actions and clarify what is important going forward. Right now, out here on my lonesome, it's simplicity, space and solitude that are helping me clarify what those might be.

/ / / / /

Early afternoon and the sun appears. Still a cold wind from the north and still fresh but a bit of sun makes all the difference. I take myself off up the Doune track and into the remote bounds of Knoydart for a few hours. I sit high up, tucked down under some rocks out of the wind and gaze out over Loch Nevis and Mallaig to the south, the Small Isles and the Skye Cuillin to the west, and north up the Sound of Sleat to Gavin Maxwell's Sandaig and Kyle Rhea, the gateway for my onward journey. What a wonderful thing it is to know a country so well, and I feel utterly at home.

Back to the tent, tea, read, doze, dinner, bed. Wind due to drop tonight and move to the south-west, so I'm planning another early start to make use of the lull if it comes.

- Route: Eilean Giubhais to just south of Doune
- Distance: 8 km (weather)
- Weather: initially grey and cold then too windy for progress
- Wind: Green/Amber. Force 3 very early morning, rising to 5 and 6 by 8am

- Sea state: lumpy and very confused on west-facing coast; wind, tide and swell playing out against each other; last 2 km were heart-in-mouth paddling
- Events: just getting here!
- Camping. Settle into more remote bay south of Doune. Long rock-strewn beach, challenging carry.

Monday 20 June, Day 18: South of Doune to north-west of Skye Bridge, 37 km. Wind: Green/Amber.

Tomorrow is the solstice. At 3am I'm wide awake and listening for the wind. It has dropped – yes! It's cold, too, and the tent is running with condensation as there is no wind to whisk it away. It is a very long carry of boat and kit over rough ground at low tide, so by the time I push off I'm already somewhat gubbed.

Great to be on calm water and heading north again. The air is cold and fresh, the sea feels warm in comparison, and a long day stretches ahead. No sun yet – but there are hints of it in a pink horizon above the mountains to the east. I have miles to go, and the Kyle Rhea narrows await.

What a beautiful place to be, the west side of Knoydart in the dawn and paddling north towards Loch Hourn and Skye. I glide through the calm water, my paddle casting droplets fore and aft, and then slowly, as the clouds dissipate (to paraphrase Coleridge), 'the glorious sun uprist'. A good south breeze blows up behind as I cross the 3 kilometres of the mouth of Loch Hourn, and then dies away again as I put in to rest at Sandaig. Gavin Maxwell's adventures with otters here have been described by many authors over the years, so there is no need for me to add to that, but I always get a wee thrill of recognition when putting ashore here – the thrill of the story and the sense of place that transport me back to the wonder I felt at reading this tale as a young boy.

The sun is now up and the scattered beaches around the skerries are glinting shell-white. Two otters come to visit, and a pod of three porpoises, all just a boat's length away and seemingly in no rush.

Paddling quietly through the skerries at low tide, I can hear seals calling, their mournful cries echoing out over the water. I remember first hearing these cries, many years ago, reverberating from deep within a sea cave out here on the west. Emanating from the dark, their long throaty moans seemed so sad, so otherworldly and eerie in the gloom. That was just my anthropomorphising, of course, and I have come to better understand them and their lifeways over the years. But still, their otherworldliness has stuck with me. These amazing creatures, at home in two dimensions – clumsy on land, fast and fluid in the water – are both knowable and unknowable at the same time. Moving between two worlds, they have understandably perhaps become bedded into our Highland folklore as 'selkies', shape-shifting creatures, beautiful and disturbing. If you could catch one on land in its human form and hide its seal skin then you might have power over it. But such power is only apparent, for the selkie will never be yours, yearning as it does for the sea to which it truly belongs.

And seals are part of the history of kayaking too. The Greenland Inuit, originators of building and hunting from kayaks, wore seal skins as waterproofs and early forms of spraydeck. Some even wore a one-piece jacket temporarily sewn to the deck of the boat to stop the water coming in. You need to be pretty darn confident in your ability to roll if you are going to sew yourself in. I like the idea of these Inuit as original selkies, too, as they shed their seal skins on coming ashore.

Curious creatures, eyes facing forwards like ours, ridiculously cute as babies, and inquisitive enough to follow a kayaker and even playfully nip at a rudder on occasion, seals are easy to love. But that has not always been the case. Seal fur and blubber was a commodity right up into the 20th century and the disturbing footage of culls on blood-spattered ice are well known. Today, here in Scotland seal culling is very tightly controlled, and for some industries, such as salmon farming, now banned altogether. It is not a perfect world, though, as these efficient hunters continue to compete with humans for fish, and should you earn your living from the sea or from sport fishing, for salmon for instance, then the colonies of seals on skerries and at river mouths can be an issue.

But that mournful cry, it gets me every time.

Beyond Sandaig the western shore of this northern end of the Sound of Sleat is precipitous, uninhabited and utterly beautiful, cloaked as

Support team

Wilson, keeping me sane…
sort of!

Inver Tote on Skye – late night sanctuary

Kilt rock, Skye

it is in natural woodlands, and alive with birdsong as I paddle by beneath, and running with water from the hills above.

For this is a wild place. Not on the land, not really, this land with all its paths and roads and crofting and sheep and managed deer. But out at sea, here there is still a wildness. Many words have been penned on what is wild and what is wilderness, and in this era, when humans appear to rule supreme, is anywhere untouched by our endeavour or our waste? Plastic bags found in the world's deepest sea trenches; whales and albatross with stomachs full of stuff we've thrown away; that seahorse clinging to the cotton bud. All of these suggest that the impact of the human species has removed the truly 'wild', the 'untouched', from this planet altogether.

It can feel hopeless until, of course, we start to take action. Then hopeless becomes hopeful and we persuade ourselves that our one drop in the ocean is – assuming others are doing the same – meaningful. But why should we care? What's the point? The problem of our waste, our emissions, our removal of native species, of biodiversity … is it not all too vast a problem? Is the planet just too big for us to care, the politics just too insurmountable? We are on a downward trajectory, are we not? – trashing our backyards to the point where the air we breathe and the food we eat is diminished and diminishing. All that is true, and some – Stephen Hawkins, Elon Musk and the like – envision a future where to survive as a species we humans will need to find other planets to live on, other planets to trash. I disagree with this pessimism, and I wish the energy put to getting to Mars could instead go to addressing some of the more pressing environmental issues here on Earth. There are spin-offs from the Mars venture I am sure, but it is the intent, the acceptance that our future is elsewhere, that seems all wrong.

So why does the seal's cry matter? Why does that lonely sound, echoing across an early-morning sea deep in the Inner Hebrides, strike such a chord? It matters because it's a reminder that there are still lives out there beyond our reach. Animals, birds, fish, mammals, reptiles, insects, molluscs, plankton, millions of creatures all continuing to find their own way without our intervention. We have intervened of course – witness the decimation of fisheries and the ubiquitously polluting plastics, for instance – but life, when given the space and freedom to do so, finds a way.

But how many of us ever hear that other-worldly mournful cry, that echo from the wild? We hear and see it on our screens, but to witness it live, with the weather coming about you … this is what makes it real, brings it alive, makes it meaningful.

And it is inspiring, too. Knowing that wildlife out there can thrive gives me hope that a future of plenty, of shared resources between humans and our fellow beings, can exist. Indeed, it is why I take myself off to these places and risk life and limb to settle to the rhythms of wind and weather and water. Henry Thoreau talks of the 'tonic of wildness' as an essential counter to urban, or what he calls 'village', life. Writing in the mid-19th century, he observed a growing industrialisation and commercialism – indeed, even the ice on his own Walden Pond in Massachusetts was commercialised for profit at one point, and his view that we should value wilderness, to have room for wonder and 'not knowing' chimes well with the modern environmental movement, and with me.

To release yourself to the sea is to give yourself a chance to realise that wildness, to feel it in all its unpredictable and restless states and, despite the many challenges that come your way, it can indeed work as a tonic. Travelling in a small boat, without electronics or the 'infernal' combustion engine, entirely under your own steam, is a humbling experience. Mother Nature rules, and on this journey that has never been more apparent. I have been kayaking these Highland waters for years, but most trips have been for a week or two, and of course many individual weekends too. For those shorter trips, my friends and I check forecasts and go to those places where it isn't blowing a gale – where it is, in other words, possible to paddle. We will generally have a car handy, parked up somewhere to return to and change direction if winds require it. That form of control is understandable; those are precious holiday times, limited in duration by work and family responsibilities, so you maximise your chances of having a good trip.

This trip, however, is different. I have removed myself from the responsibility of work for a while, saved up the cash, taken the risk of a break in career and cast myself forth on the sea. The journey I want needs to be long enough for me to have time to flex and change as nature dictates and to take the rough with the smooth rather than rushing back for the car every time the wind changes. It is a rare thing to have such time and mental space to undertake such a venture, and I find myself rediscovering these remote Highland seaways anew.

And it is remote. And in exposing myself to these hard and often barren coastlines I have begun to come out of my own head at last, to see myself in my kayak from afar – tiny and insignificant in this vast space, just one other being amongst many. I paddle through quietly, I hope, leaving nothing but a slight wake that disappears in a moment, and I look, observe, think, wonder and hopefully learn the lessons nature provides. I often find myself gasping with awe at a vista, a change of light, a dolphin encounter, a running sea.

Beyond the immediate reach of humans, it's the natural world and perhaps especially the sea, that seem to satisfy some innate need in us. We seek out the sea, do we not, on our holidays for instance, and we seem drawn to open spaces, to trees, to rivers. Something in the natural space draws us in in a way that the built environment does not. We know already that our minds work more efficiently, more creatively, when we are exposed to natural things. Darwin, Einstein and others famously did their best thinking outside. Annie Murphy Paul, whose 'extended mind' work I mentioned earlier, points out that we did, after all, evolve outside over hundreds of thousands of years, and that legacy remains buried deep in our brains. The multiplicity of shapes, colours and movements that make up the natural environment are exponentially more complex than any modern structured town or city with its evenly spaced streets, buildings and windows. And yet the corners of the brain associated with stress do not fire up to make sense of all that natural complexity. Indeed, they do the opposite; they settle, they calm. Nature makes us calm. And in the calm, thoughts can curl about each other, reform, make new sense of things, learn, create.

I can feel my own brain settling, and the awe I feel cements in me a comfortable knowing that there is indeed a world out here beyond human limits, a world that will recover if we do ourselves in – or, indeed, disappear off to Mars and beyond. The mountains, rocky coasts, beaches, headlands, inlets, seaweed-covered shores, bird-spattered cliffs, moaning seals, soaring eagles, diving gannets, crying loons – all these need us not. And then there is the sea. Never still, always changing in tone, in colour, in steepness, as fickle as the wind, uncaring, unknowing; just 'being'. This is the wild place I have rediscovered. Forces limitless, vast and vigorous, envelop my tiny person and leave me sometimes elated, sometimes fearful, always respectful – and, yes, often in awe.

/ / / / /

The Kyle Rhea narrows quickly swing into view, and I start to hear the gentle splashing of moving water over the shallows. The rising tide is still running, and as it takes me into its grasp the land begins to speed by. A few lonely-looking houses on the western shore are quiet – it's still early after all – and I seemingly have this iconic place all to myself. The tide runs fast here – it's a tight pinch for all that water from the Sound of Sleat – and as I speed through, hardly paddling, there are boils and eddy lines springing up and dying away again all around. Seals are hunting in the flow, and another otter appears at the northern end. They are here for the fishing, of course, and I imagine the flashing bodies and slashing teeth of the feeding frenzy below. They are so totally at home in this powerful flow, and in comparison I feel small and inadequate, balancing precariously on top while all the action is beneath.

My timing is perfect for once, and as soon as I get through the narrows the wind freshens from the south as expected and the spray from my paddle is lifted up over my head. A fisherman, his boat heading the opposite way, into Kyle Rhea, passes close by, waves from his cabin then comes out to lift his arms and eyes to the sun in celebration of the return of the light. I reply in kind, and we share a moment of recognition, in passing, as two seafaring folk welcoming the return of the sun after so much dreich.

I have been going now for four hours, so I stop to rest south of Kyleakin on Skye and give Leah a call. She answers quickly, and we decide to meet up in Portree at the end of tomorrow. I am paddling there of course. She is keen to come out and support, and will bring some clean clothes, more food and more tent poles – another one broke yesterday – and I love that we are able to share this adventure. It means Leah feels involved in a way that had I disappeared off to Norway she would not.

On then to Kyleakin; no stopping, wind rising fast as forecast; under the Skye Bridge at the very top of the tide, and I can see the sea

Dawn at Staffin, a moment of calm

Sanctuary from the wind at Port Gobhlaig, Skye

Dawn calm at Rubha Hunish, gateway to the west

out in what is known as the Inner Sound north of Broadford whipping up white. The blow is now round to the west-south-west and gusting hard. It is these gusts that can knock you over, so I keep a weather eye. Round the enormous pier at Mowi's salmon feed plant, and then it's head down and grinding it out against the wind for a few kilometres. A small rise to the west of an even smaller beach offers just enough shelter for me to risk setting up the tent. As I don't have any more spare poles I add additional guylines and take some time cutting holes and threading cord to create a veritable spaghetti junction of support. This tent ain't going nowhere! It'd better not, as it's a windy and wet night forecast.

- Route: South of Doune to 2 km west of the Skye Bridge
- Distance: 37 km
- Weather: calm, then windy and gusting from south-east, then round to westerly on the nose
- Wind: Green/Amber. Force 2 to 4
- Sea state: calm to windy after the bridge
- Events: otters, porpoise by the boat, sunrise, crossing Loch Hourn in the dawn light, Kyle Rhea narrows, fisherman with arms aloft
- Camping: tucked under a tiny hill. Wet night to come.

Tuesday 21 June, Day 19: Skye Bridge to Portree, 37 km. Wind: Green.

Calm. Mirror-calm. I have a long day ahead, to get to Portree to meet Leah, and after a wind-blasted night the calm is a welcome distraction from feeling permanently put upon by the weather.

Funny things happen in the calm. The sea feels like treacle, and as I'm paddling towards the Isle of Scalpay on the low tide the perspective also makes it feel like I'm paddling uphill. Something to do with the mountains looming ahead and the pinch of the view into the island narrows, perhaps. Pushing these negative emotions aside, I settle to a rhythm and let my mind drift; with no risk of capsize you can enter a sort of zen state when the movement of body and paddle and boat synchronise without conscious thought. Arms held high enough to get a decent speed but not so high that you are over-straining; waist twisting, core engaging, paddle dipping and rising on repeat, mind settling. As the bow cuts the water, slanted ripples flow out in both directions leaving a watery V that lasts for seconds and then fades, leaving no trace.

High overhead, a large skein of geese, faintly honking their way north, mirrors my V in the grey skies above.

I love entering this state, this inner calm when the day's events and any troubled thoughts settle out and disappear. Actually, I don't really notice when I'm entering it and only realise it's happened when it passes. On the yoga mat at home, during a *vinyasa* flow – that wonderful fluid transition from one yoga posture to another in time with the breath – my mind slips away and I can easily lose a couple of hours. Except, of course, it is not a loss; it is a gain. A positive thing, akin perhaps to waking from a deep sleep when swirling thoughts and breath and heartbeat settle. Here, on this floating mirror, paddle stroking rhythmically through the water, I find that time passes without my noticing. Calm sea, calm mind. Around me, the towering landscape of eastern Skye looms then falls astern, and the open coastline of the huge Trotternish Peninsula rises to the north.

As I criss-cross Raasay Sound, memories slip in and out of mind as I paddle north: a friend's wedding on Raasay, wind-driven kayaking trips, and even once a coastguard call-out north of Portree. At Camustianavaig the coastline steepens, and the cliffs and steep ground of Ben Tianavaig hang high off my port side. I look for sea eagles, knowing they're here, but today they must have business elsewhere. At sea level, huge boulders fallen from the hill have settled in the shallows to create underwater rock gardens. Changing with every state of the tide, this ground would be difficult for paddling in any onshore wind, but today it's a calm pallet of blues and greys, softened by a living mass of gently waving kelp and dotted with the bright pink of sea urchins clinging somehow to the rocky slopes.

Seven kilometres on, at the turn into Portree Bay, the Trotternish Peninsula looms large. Its spine is a 25-kilometre ridge with cliffs extending the length of its eastern edge. Fabulous bits of geology such as the Quiraing and the Old Man of Storr draw visitors from around the world, and the age-old dilemma of landscape protection versus access versus tourism is there for anyone with the eyes to see. Far away from where I am sat at sea level, I know there are cars parked on verges, campervans clogging single-track roads, tempers rising and tourists ticking boxes. Money oils the wheels of these tensions, though, and goes to support local business and community, which is good.

But I'm hopeful that it's not just about money, and that when visitors view these places something else happens. Hopeful indeed that the places are more than just a tick in the 'to see' box, I like to think that they might also excite some wonder, some awe. To walk out to the Quiraing is to journey deep into time and, if you're open to it, into the imagination. Everything here is on the move. The pinnacles and tabletops of rock, including the stack of the Old Man of Storr to the south, are the result of land slipping back towards the sea. Geological forces, epic in scale, having lifted the rocks so many millions of years ago, are now, with the help of wind, water, and ice, undoing their earlier work. Cracking cliffs, falling boulders, stacks and sliding slopes are witness to this eternal change, and to walk amongst it, or indeed to paddle beneath it all, is to feel humbled and small.

When I turn west into Portree Bay, Leah meets me at the jetty to the east of the town, and we spend the rest of the day sorting kit, drying clothes, mending tent poles, topping up food stores and catching up with news from home. The B&B is simple – but compared to the camping, utter luxury. The owner's son, Andrew, comes to say hi. He's a leading Highland road runner and came second in this year's Skye half-marathon, pipped to the post by Gordie Lennox, an old charity colleague of mine, who has become one of the Highland's best distance runners. My claim to fame is that I introduced Gordie to mountain-running in the rough ground around Loch Luichart some 15 years ago. If we went running together now all I'd see is his heels, for sure.

- · Route: just west of Skye Bridge to Portree, via Raasay
- · Distance: 37 km
- · Weather: calm, cloudy
- · Wind: Green. Force 2
- · Sea state: mostly calm
- · Events: zen
- · Camping: B&B.

Wednesday 22 June, Day 20: Portree to Invertote, 19 km. Wind: Amber.

OMG, what a time of it today. Set off from Portree mid-afternoon in a growing south-westerly wind in the expectation that by the time the wind grew in strength as forecast, I'd be tucked in under the precipitous easterly shore of Trotternish and heading north. The longer-term plan is to get up to the top of the peninsula within a couple of days and take advantage of the one day of calm forecast for 24 June.

So much for working my way into a sheltered shore. It turns out that the south-westerly is bent round the corner by the pinch of Trotternish and Raasay offshore, so I find myself paddling a Force 6 on the beam and slightly aft. It is the noise of it and the feeling of utter exposure that I'll remember most. Just 100 metres out and all the way across the Sound of Raasay the sea is white with breaking waves, and close in, the shallows raise it up a notch – and the submerged rocks on this coastline, so beautiful in yesterday's rock gardens, now become a hazard. I'm probably fairly flying along, but I feel very alone and on the point of losing it many times. It is the gusts that do you in, disrupting the flow and taking you by surprise if you're not checking over your shoulder all the time.

I am not sure how long I bounce around out there but, frazzled, I get to the only relative shelter since Portree of Bearreraig Bay, 12 kilometres on. Big bouldery shore, lumpy grass and, unexpectedly, a small hydro scheme on the south side. There is no vehicle access here, just a steep 200-metre narrow-gauge winch railway for getting equipment up and down the cliff to the hydro building itself. No people. Wind and cold, rain coming and going, overwrought from being overexposed in uncomfortable seas, I put up the tent – and the gusts promptly break another two poles. I need to wait out this wind and hope it will dissipate this evening. Even my ever-smiling friend Wilson's natural optimism has left me cold today.

A walk along the shore to take my mind off things leads to a fossil ammonite, 30 centimetres long and clear as day, raised up on its rock. This part of Skye is already well known for its Jurassic fossils, and as well as the dinosaur footprints in the Staffin area, the more recent discovery of a 2.5-metre pterosaur has put Skye firmly on the fossil hunters' map.

I suspect the sheer size of my ammonite finding has saved it from being carried off, and I hope others will leave it in situ too.

By 8pm, gazing out on the sound and watching a pod of six Risso's dolphins porpoising across the wind-blown bay, I persuade myself that the wind and sea have dropped a notch. Steeling myself, I pack up kit and boat once again, haul it back down over the slippery boulders and set off. The moment I leave the northern end of the bay, the sea picks back up and I'm burled along all the way to Invertote Bay, 8 kilometres away, somehow without any capsize. Same sea, same gusts, same frazzle, same self-talk, mostly around 'what the hell am I doing out here!'

Ashore at Invertote Bay, more of a shallow scrape than a bay really, and tucked under the cliffs with the wind whistling about and the sea noisy on the boulder-strewn shore; it's 10.30pm and I am just into the tent. Need to sleep.

- · Route: Portree to Invertote Bay
- · Distance: 19 km
- · Weather: cloudy, overcast
- · Wind: Amber. Force 4 to 6 from south-west
- · Sea state: biggest and most challenging yet, wind-blown, uncomfortable, exposed
- · Events: pod of Risso's dolphins porpoising across Bear-reraig Bay; fossil
- · Camping: grassy, wet.

Thursday 23 June, Day 21: Invertote to Gobhlaig Bay, 21 km. Wind: Amber/Red

Another early one to catch another potential weather window. But again, that does not materialise and the 2.5 kilometres to the Culnacnoc turn is as gnarly as anything yesterday. At Culnacnoc the coastline first curls east then takes a turn slightly north-west, creating what I hope will be some shelter from this wind-blown sea. Last night was a short one. I didn't get camped until after 10pm, and by the time I had fed and sorted kit it was well past 11pm – and I was planning this early start. I poured with sweat all night, too. It's stress, I think, and all self-imposed, of course. I'm in a very exposed position here with tricky seas, high winds and few get-outs. And I'm on my own. It's the latter, in these winds and seas, that turns up the heat.

Sure enough, though, at Culnacnoc there's a moment of relief. And it *is* just a moment, really. The pleated granite folds of Kilt Rock and its associated waterfall and extensive cliffs lie ahead, but only a kilometre north of the headland the same lumpy sea reappears, curving round the cliffs. Powerful gusts, occasional breaking waves, rock gardens lumping the water into a confused clapotis, and once more that lonesome feeling that if something goes wrong I'm on my own. My self-talk holds me to the task, and I take temporary shelter in amongst some of the rock gardens wherever I can. An amazing feeling, sitting there, bouncing about between the small stacks and offshore slabs with the messy sea all about. I feel like I'm being preyed upon, hiding out while the wind and water howl and grab at me through the rocky walls. The wind has risen, and the gusts that I don't always see coming are really unsettling. I have to brace continually today, and the worst of the gusts seem to be dropping down off the cliffs. These downdraughts flatten the water into a wide circle and then spin out in all directions, making this kayaker nervous as hell.

I get through, though, shooting several tight gaps on surfing waves, missing several rocks by inches and looking up only occasionally, just enough to get a glimpse of the seabirds in the cliffs, the waterfall from Loch Mealt getting partially blown back skywards by the gusting wind and a glimpse of Kilt Rock itself. Paddling beneath it in this weather, while scaring the hell out of me, is a thrill too. Its curving basalt columns, less broken than those on either side, do indeed mirror the folded cleats of a kilt. But impressive as it is, it's the waterfall being blown back uphill, the birds buffeted about the cliffs and that *wind* that dominate my day.

The coastline continues to bend north-west and with it the sea slowly becomes less messy, leaving only the gusty, fluky wind for me to deal with.

Staffin is a welcome site. As I turn the corner to Staffin Island, instant calm and a wave of relief sweep through me. It's only 6.30am, grey and cold, yet with two hours of an intense day already under my keel I feel on edge and a bit jangled. No one's about, so I put ashore, brew up to try

Luskentyre, Isle of Harris – a hard won idyll

Luskentyre, Isle of Harris – a dry camp at last

Luskentyre perfection

and give myself some cheer and then, as so often happens when you're out there all day, things change around you. Emerging from behind the grey, the sun breaks free and flushes my melancholy away. What a tonic, what a reward! I pull out all my wet gear and drape it over rocks and sand to dry. The sea's sparkling blue and white with wind, Wilson's smiling, and I'm on target to catch the weather window for tomorrow to get round Trotternish. Suddenly the world seems right again.

Revictualled and dried off, nerves settling, I set off for Port Gobhlaig in Kilmaluig Bay, the tiny community perched on the tip of the Trotternish Peninsula. Caves, an arch, rock gardens, basalt columns, seabirds and of course wind, ever more wind, fill the journey with interest and challenge. Somehow once again I manage to stay upright in amongst the gusts.

Port Gobhlaig's few houses are barely sheltered from any winds, really. Westerlies and southerlies rush down from the hills to the west or up the sound, easterlies batter in straight from the sea, and northerlies pile the water up into the bay. It is far from the perfect place for a community reliant on fishing, for there will be many days when you just cannot get a boat out. But back in the day it was indeed the fishing that brought settlers here: quick access to fishing grounds, while others had long sea journeys to reach these (then) fish-rich waters. First boat out is first to catch.

I make landing at low tide in a very sloppy seaweed-covered cut in the rock. Haul everything up, find an idyllic spot to pitch the tent, put my head down and fall into a deep sleep.

The rest of the day is spent sleeping, reading, eating and walking out to the point which I need to round tomorrow for this journey to continue. From my hilltop lookout, Rubha Hunish looks horrendous today. It is so fierce a turning point that it has been named twice; *rubha* is Gaelic for 'headland' and *hunish* is Norse for the same. As this point pokes north out into the Minch, all that water running north or south on the tide must curve around it, and of course it speeds up as it does so. Rubha Hunish is a place to respect and to plan your way through. Today, at peak tidal flow, with wind blowing in the opposite direction, there are standing waves breaking for a couple of hundred metres off the point. Tomorrow, in the forecast calm, I'm aiming to get there at the top of the tide, and by then it should hopefully be flat.

- · Route: Invertote Bay to Port Gobhlaig
- · Distance: 21 km
- · Weather: overcast then sun
- · Wind: Amber/Red. Loads of it; forecast 4 to 6 and gusts on top; really challenging
- · Sea state: busy, confused, wind-blown; calm at Staffin but windy again as moved north
- · Events: the paddling, the wind, sun appearing at Staffin, the walk to the point where a fox flowed russet red over the hill, sleeping
- · Camping: at village edge.

Friday 24 June, Day 22: Port Gobhlaig to Uig, Ferry to East Loch Tarbert, portage to West Loch Tarbert, then paddle to Luskentyre, 36 km. Wind: Green/Amber.

How much variety can you pack into one day? Here it is:

Kayak on the water by 4.30am to catch the weather window and tide; manage to get the tide exactly right round Rubha Hunish, which is, I feel – rather irrationally – disappointingly calm after all the hype; stop for porridge at old pier en route to the ferry port at Uig; reach Uig by 9.30am, sort ferry ticket for Harris, portage all kit and boat from landing spot to the end of the very long jetty; tea shop; get cheeky and ask two Swiss lassies with a VW Camper van if they'll help me get my kit to West Loch Tarbert from the ferry, and they say 'yes' (ah, the kindness of strangers once more!); ferry; liaise with *les Suisses* at the other end to move the kit; use the trolley to take the boat over the half-mile portage; paddle 10 kilometres with wind on the nose; swell curling round Taransay and rising in the shallows, choss,[5] clapotis, not for the faint-hearted; relief of rounding the corner to Luskentyre and to a familiar view; surf landing; long carry of kit up the beach as now low tide; sun-kissed camp spot in the dunes, machair flowers bobbing in the breeze; ready the tent for the strong winds coming tomorrow; swim, wash, feed; sun going down over Taransay; sleep to sound of surf.

---

5 'Choss' is a climbing term, really, that describes loose and unreliable rock. However, many paddlers use it too, to describe messy, somewhat chaotic sea conditions.

Not bad eh? I'm here in the Outer Hebrides at last. The plan has worked. The wind is due to rise from the east tomorrow, so I've put myself on the west side in the hope I can make a bit of progress south.

- Route: Port Gobhlaig to Uig, ferry to Tarbert on Harris, portage across to west side, paddle Tarbert to Luskentyre
- Distance: 36 km
- Weather: calm on Skye, then wind, chop and swell on Harris
- Wind: Green/Amber. Ditto above, Force 2 to 4
- Sea state: ditto above again
- Events: early-morning paddle, breakfast on old pier, meeting *les Suisses*, the lumpy Tarbert to Luskentyre paddle
- Camping: exquisite spot on Luskentyre.

4 | Surf-ridden

Days 23 to 28 – Luskentyre to Lochmaddy

Saturday 25 June, Day 23: Luskentyre to Taigh Bhuirg, 8 km. Wind: Amber/Red

Supposed to be a day of rest today, and kind of is and isn't. Day dawns gorgeous, sun up and deep black clouds to the west, but not forecast to come this way. Yoga on the spectacular Luskentyre Beach on my own: early morning, turquoise sea, white beach, a good breeze but not cold, pounding surf and the flashing gleam of gannets hunting out in the bay to complete the idyll. Taransay, and the hills of Harris to the north, are turning from blue to green as the sun strengthens. Wind due to blow Force 5 to 7 this morning.

Surf is coming in on patterns which are discernible only by standing still and counting waves for long enough. An apparent short-term pattern – five large breakers, then a lull – dissolves the longer you observe. There *is* a pattern, though, and it seems that the biggest waves, the ones that show vertical blue before collapsing under their own weight and momentum, come in groups of seven. Then there is a wee lull, still surf but acceptable from this kayaker's eye view. It is an intimidating prospect to take it on, and I do not pretend to be particularly well practised – but wimping out means I won't make any progress. So I'm standing here staring down the waves, looking for patterns and ways of getting through without capsizing. Getting it wrong means not just a wetting – I'm used to that – but the potential for serious injury. Capsizing with a fully laden boat risks the surf thumping

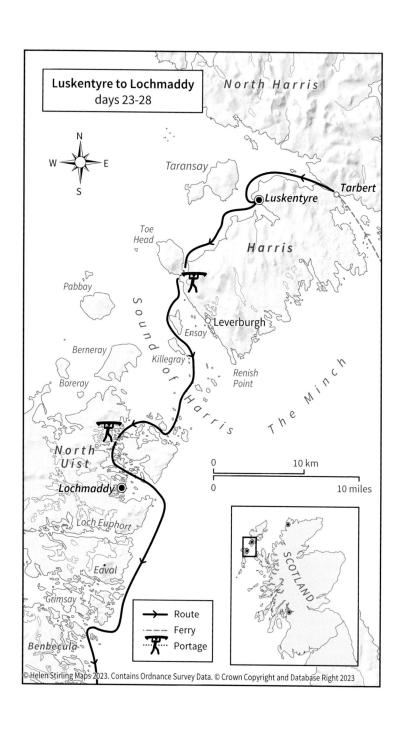

Luskentyre to Lochmaddy
days 23-28

N
W — E
S

North Harris

Taransay

Tarbert

Luskentyre

Harris

Toe
Head

Pabbay

Leverburgh

Sound of Harris

Ensay

Berneray

Killegray

Boreray

Renish
Point

The Minch

North
Uist

Lochmaddy

Loch Euphort

Eaval

Grimsay

0   10 km
0   10 miles

SCOTLAND

Benbecula

→ Route
--- Ferry
Portage

© Helen Stirling Maps 2023. Contains Ordnance Survey Data. © Crown Copyright and Database Right 2023

you upside down onto the beach with all the weight of the boat landing on your head; that's a broken neck! So I take these things on with some thought, I hope, and when you are on the edge of your abilities and experience, it is a marginal decision by nature. Too little caution and the body breaks, too much and you'd never go anywhere. The trick is to find the balance, use your experience, observe carefully, and then of course make a decision and commit. These are the thought processes, 'ways of being' if you like, that add endlessly to one's abilities and expectations. Sometimes I fail, sometimes I do get wet, hopefully I'll never break my neck … and then, as I punch through the surf, whooping with elation, I add another 'knowing' to the bank of self-knowledge.

I think that at 58 I know what thrills me and what terrifies. I don't lust after the scary roller-coaster ride or the white-water river or the bungee jump. I'm not a great fan of heights, either. Being in the mountains and even climbing are fine – but jumping off stuff from a height, that's not a thrill, that's a kind of horror. I've done plenty of it, not least in the Royal Marines, but also with a bit of parachuting, climbing, river kayaking and coasteering thrown in. But, as I stand on the edge, where others appear full of excitement, I'm quietly hiding my terror as my stomach tightens and every fibre of my being says 'No!' Sometimes, though, you must face these things. Sometimes it is because you're in the company of others, a kind of group pressure, and sometimes I'll do it just to test myself.

/ / / / /

But standing here now, the surf crashing on to the beach in front of me, a heavily laden kayak at the ready, it's clear that without the desire for the journey ahead I would likely be heading straight back to the tent. But ahead lies the next horizon, new seas, new islands, as yet unknown experiences, and just more of this place that has such a hold on me. So out I go.

Holding the boat in shallow water, trying to keep it pointing into the incoming waves as the collapsed ones surge around my knees, counting the breaks, making the decision, leaping into the boat, spraydeck not fully on, PLF into the lull I'm hoping for, and realise immediately that I've got it wrong. The first break washes over the deck and into the cockpit, and the dark intimidating wall of the second rears up ahead. My heart stops, a moment of fear, then I think 'just *paddle* it'. A couple of hard strokes, bow pointing skywards, wave breaking, water rushing, a physical punch to the chest, my head takes a soaking – and I'm through!

Reading back on this piece I wonder if I am being melodramatic here, overreacting perhaps. I don't think so. I'm being honest, really – and what is a journal if you can't share your inner thoughts and fears? The sheer physical reality of journeying by kayak alone through these seas creates an intensity to life, to the everyday, which just doesn't exist in the same way on land. Maybe this is the word that best captures this new way of living: intensity. Another would be 'resilience', but 'intensity' better describes the immediacy of the moment and captures both the fear and elation in equal measure. And in this journey there have already been many such moments: breaking through a surf line for instance; the dawn light when alone on the water at 4am; the massive swells dwarfing the boat; the fierce squalls and the tricky seas on the stern quarter; the realisation that I'm going to be exposed to a big sea for a few headlands yet; and the sun breaking from a rainy day. And then there are the visitations: otters, porpoises, dolphins and sea eagles; the skeins of geese honking patterns across the skies silhouetted in the morning sun; the gangs of guillemots heading out to fish in deeper waters; and the folded arrows of the gannets' reckless plunge. And of course the backdrop of it all, the mountains, cliffs, islands and skerries that frame the whole. Within all these are my moments of intensity, and even now, over 40 years after discovering the Highlands for the first time, even now after exploring so much of this coastline, still they set my heart beating.

'Set my heart beating' still doesn't do justice to the feeling, really. 'Elation' would be more appropriate. Science would tell me I'm getting an endorphin rush, a chemical flush that we have evolved to enjoy so we keep doing the things that are good for us, activities that are pleasurable such as eating, running, socialising and sex. But we are more than biological animals, I think. We are emotional and spiritual ones too. We have evolved culturally as well as biologically, and encultured as we are into our birth societies we cannot help but respond to their ideas implanted early in life.

The evolutionary biologist Richard Dawkins called these cultural genes 'memes'. I like that. Born with an instinctive desire to thrive, we learn to survive not just our physical environments but our cultural ones too. These are different, but not dissimilar, in all societies across the planet, and they implant ideas and ways of being that are difficult to ignore. Some, such as manners, kindness, determination, sociability, empathy, are good for us, while others, such as violence, envy, deceit, laziness and the like, are – at first glance at least – not.

To personalise this still further and perhaps discover where some of these feelings of elation come from, I'll share some of my early background. Having been sent away to boarding school at the age of nine, my enculturation included a strong and somewhat authoritarian religious element from primary right through to secondary school. This was manifest in two ways. First in the regular church-going with associated singing, praying and sermonising, all of which were collective enculturating experiences. And second, in the value system by which all us boys lived. For, yes, it was all boys up to the age of 16. Don't get me wrong – I met lots of girls, too, especially in the large group of my friends back home who in holiday times met and snogged and danced and played sport and fumbled our way to sexual maturity. But at school all was values, responsibilities and consequences. The latter, at primary school level, was felt literally, on the occasional beating administered by the headmaster when we got caught dissing those values. Six of the best on the arse with a walking stick told us we'd done wrong and who was the boss. The teachers, especially head teachers and heads of houses, ruled. We obeyed, rebelled occasionally, got punished. And repeat. And all the time those cultural values of working hard, playing hard, respecting authority, respecting God, seeped in and became us.

So why am I revisiting all that stuff? What has it got to do with these intense feelings of elation and calm experienced out here on the remote Atlantic edge? Well, I think I've come to accept that I have a mostly unconscious encultured respect for authority, ultimately manifest in God perhaps, and as a result maybe a rather too deep-seated need for said authority to keep me safe, to keep me right. My thinking is poorly defined, but there is something there that pops up in every intense moment. I often find myself offering a quiet, not always unspoken 'thanks' to something godlike as I come through yet another necky sea or a prolonged exposure to fear. I don't do it intentionally – indeed I'm not a religious person, or at least I don't think of myself as one. I don't attend Church – perhaps I had too much of it forced on me during childhood – and I'm uncomfortable with the structure, the hierarchy, the occasional self-righteousness, the certainty, the duty. All that. But still there is something in me that accepts one can be both rational and spiritual at the same time. Yes, I could rationally explain my elation, those intense moments of feeling looked after, using biology and chemistry but – you know what? – it doesn't satisfy. What satisfies, in the moment, is recognising, accepting, that there are forces out there: wind, seas, mountain-building, seasons, tides, sunshine and more which dwarf us; and compared to a short human life – indeed, compared to the blink of a geological eye that is the entire human existence – these powers of creation, natural or otherwise depending on your view, are something of a 'wonder'.

As I get older I find I'm less afraid to wonder, to express an occasional 'wow', to accept that some things, some happenings, are just awesome in themselves. The heart-rending beauty that appears when a sun sets all aglow, when the gannet dives, when kittiwakes waterfall off the cliff, when the storm surges up the same, when the sun appears after rain, when that perfect mix of sea, island and mountain is captured beneath a rainbow. Is that God speaking? I don't know. I doubt it, but it is without doubt the result of eons of time, of endless change, and I am the statistical miracle there to see it at that moment. It feels like a gift and I feel the need to give thanks.

I mentioned John Muir earlier in this journey. His musings on his extraordinary travels across America matched an investigative mind, seeking to reconcile the geological processes that created Half Dome and the wider Yosemite landscape, with a 'wonder' at God's creation. He too was enculturated into a God-fearing world, and could not help himself, it seems, from finding pleasure in both the scientific and the spiritual endeavour. That is precisely where I am, I think.

/ / / / /

So here I am, through the surf, wet but on my way. The journey continues, momentum restored, addiction to the horizon satisfied for the moment.

But what a flippin' horizon is in wait! Across Luskentyre, hauling my way into the easterly blowing hard across the bay, into a moment of relative calm under the far shore, round a headland, and there quite a different sea lies in wait. A large swell has built somewhere far out at sea and is curling round the western end of Taransay, rising up on the ever-shallowing coast and dumping a huge surf everywhere I look. Thankfully the easterly breeze is dropping slightly, so the wavy choss on top of the swell is just about manageable. But still, it's an intimidating place. I am now well out from the shore, keeping clear of surf lines off the white beaches and clapotis-ridden shallows. Tiny figures on distant beaches just add to my growing panic as I realise I am out of my depth here. Landing in this dumping surf would be crazy. I stay clear of it, leaning on the water with my paddle strokes, heading slowly west, riding the big old swell and trying to keep calm.

Where to land? Everything is awash. A small very rocky shore, slightly east-facing, offers a possibility, but it's not great. I paddle on. There is another beach on my map a kilometre or so further west. Slowly it comes into view, but like the previous ones it's awash with dumping surf. Now I start to panic a little and recognise a new sensation, something deeper in the gut, which I think is fear. I have to keep going – but where to? Options are disappearing. And, at that moment, just as I'm passing the western end of the dumping beach, there is a line of rock exposed by the low tide which reveals a tiny cleft and a finger of sand just a few metres wide with only a light surf. Who would have thought it? My tiny heart lifts and I surf in, hit the sand and am safe. Wet, frazzled, but safe. Such intensity in so short a time.

The only place fit for a tent is a slope high above the shore, so I have some clambering and carrying to do, and it's past 10pm before I'm inside the tent. Water on, tea, rice and tuna, and sleep with the sound of the surf.

- Route: Luskentyre to tiny headland at Taigh Bhuirgh
- Distance: 8 km, but feels a whole lot further
- Weather: sunshine, very windy
- Wind: Amber/Red. Force 5 to 7 dropping to 3 to 4 early evening

- Sea state: all about the surf, swell and wind-blown sea today
- Events: yoga on Luskentyre. Deciding to break out early evening; breaking through the surf; big old seas west of Luskentyre, pounding surf, fear
- Camping: grassy headland above the rocks. Views west to east and the sound of pounding surf all night.

Sunday 26 June, Day 24: Taigh Bhuirgh to Traigh na Cleabhaig (Toe Head), 8 km. Wind: Amber/Green.

A wet night, and the reinforced guylines seem to have done their job as I have no broken poles for once. At 4.30am a peek out the tent tells me I'm not going anywhere for a while; the surf is still dumping on the beach below, and the big old westerly swell is lifting the sea just as it was yesterday. Ahead lies Toe Head, a massive, bulky peninsula sticking out west into the Atlantic. It is a place I have rounded a couple of times in previous years, but right now it looks very intimidating. The heaving swell will be dumping itself onto the peninsula's rocky northern and western sides, and the seas at the point itself would be scary stuff. I decide to see what the day brings, and if the wind and swell begin to lessen later I'll go for it then.

I am running out of water so take a damp tramp to a burn a kilometre or so away across the band of lumpy machair between sea and hills to the east. Choosing your source of water is not complicated here in the Highlands – there is plenty of it after all – but getting it right is important. No one wants giardia or other bugs from drinking water contaminated with animal faeces or dead beasts, and I choose not to filter but instead to boil most of the time – okay, some of the time – well, okay … hardly ever, to be really honest. The water off our Scottish hills is usually not long from the sky, and away from habitation or farmed animals is clean, pure and delicious. Low down on the hill, the burns have filtered their contents through peat and a succession of pools and falls. The brown peat residue filters itself out, and you can drink deeply.

As I spend more time outside in the Highlands, the immersion in and recognition of our natural capital is constantly enhanced. Water – fresh, clean, naturally filtered, provenance known, without human intervention and without chemical inclusion – suffices, I find. Why wouldn't it? There's a lovely feeling of elemental oneness as you dip your head to the pool and draw deeply straight from nature untouched. Don't get me wrong, I love a cup of caffeine too, but think about it; your coffee, your tea, your infusion of camomile and peppermint, whatever your tipple, comes with baggage. It has travelled miles, changed landscapes, enslaved thousands, built empires, started wars and added carbon in its journey to your cup. A drink from the hills in our backyard, by comparison, is a guiltless pleasure to savour.

/ / / / /

Much of today is spent reading, writing and listening to the ever-pounding surf on the beach below. Occasionally above the booming, whenever there is a brightening in the sky and the rain eases, a skylark raises its voice and the world feels sunny again.

At some point, mid-afternoon and contrary to the forecast, I sense a dropping of the wind and sea. The white waves give way to grey, and while it still looks lumpy out there some of the sting has come off it. Is this a lull? Will it last? The forecast says *no, it won't*, but I'm impatient to move, to take these wee windows of opportunity when they come. It looks, as ever on this trip, marginal. The wind could drop or get up again, the sea looks like it's fidgeting between windy/scary and breezy/okay. It is rare on this journey so far that I've had an easy decision to make. There have been a few calm days, especially in that first week, but those have been the exception.

I am trying to move west and south towards Lochmaddy, and although I'm in the shelter of a westerly coast with an easterly wind what's making things tricky is the mix of strong easterlies whistling through the gaps in the hills at Luskentyre and Scarista Bay when they come up against this big old dumping swell from the west.

Marginal decision made, I undo all the graft from last night and pack up a wet tent, then lug everything down the rocks and onto the sand. This is standard daily procedure for me now, lumping stuff about, sometimes more than twice a day if the weather changes. It takes about an hour each time, and it's not the packing but the carry of kit and boat to the water's edge that can be so tiring. If it's raining then all the drybags get stowed wet. If I was careless and ripped them, all my clothes would get soaked. That would be a disaster, especially out here in this weather. Getting warm and dry at the end of a day is important. I'm okay being damp – indeed it's impossible to avoid it when the air is this saturated with water – but wet, properly wet, is life-threatening and to be avoided. So I take care, check and double-check what I'm doing, move carefully and keep an ever-watchful eye skyward to gauge when the next squall will arrive.

This tiny bit of sand, sandwiched by rocky walls and with a surging booming surf just a few metres away, feels a little put-upon. The surf on the other side of the rocky spur seems higher and further up the beach than in this tiny cove, and as the tide begins to rise the surge spills water over the rocks and onto my side. It's a race onto the water, as ever.

Boat packed, glugging water to counter the sweated effort of the last hour, drag the kayak into the shallow waves, wait for the right moment, legs in, spraydeck snapped shut and PLF to get out. And breathe!

I check Wilson's okay – he seems happy – and head west. Round the first rocky outcrop, and yes, sure enough, the swell is huge. So hard to tell from the land, but now, from my vantage point seated low in the boat, the top of the swell coming my way looks and feels intimidating. The world behind the oncoming wave disappears as I sit in each trough, and then reappears as I ride the crest. This is the same south-westerly swell I felt yesterday, but now seems to get bigger the further out from the shadow of Taransay I paddle. The open Atlantic come to visit.

I paddle hard to get out from the surf-ridden coast, away from all that breaking power. This is the same power, the same energy that shatters the cliffs, that rounds the boulders on beaches and that powders the uncountable billions of shelled molluscs to create those white west coast beaches we all love. It's early evening, and with the sun hidden by dreich all is grey and overcast, rain showers are piling through, and there are no tiny figures on beaches looking out to see me capsize should it happen. It must not.

For this journey to continue I need to get round Toe Head. It's another turning point beyond which new possibilities open up. But here now, riding this huge swell, wind blowing the water into a confusing jumble – and an offshore wind at that – I start to feel really vulnerable. There will be no let-up at the point, indeed it's likely to be much worse, especially on the windward side where the energy in the swell will be bouncing back off the cliffs.

I realise I'm not breathing; I'm tensing up. I try to focus on the self-talk, but the sea, the exposure, the sheer 'aloneness' of the situation and with worse to come, mean it's not really working. Surely this is just mind over matter. But I think this is fear creeping in. I am fearful. I am at least a kilometre out from the shore, and a breaking shore at that, and am getting lifted by a big old swell. I can handle the wind as it is at present, but the forecast is for more, and there is at least another hour of this and worse before any potential shelter appears. I try not to acknowledge the fear for a bit and just keep paddling towards Toe Head, babbling my trusty self-talk. But this time it's not working, and a really uncomfortable self-doubt is creeping in with, once again, the stirrings of panic. Kayaking is not like biking or running when if you don't like it you can just stop. Out here you're fully exposed, committed – and in my case, today, right on the edge of what I'm capable of and I feel very, very uncomfortable.

So, make a plan, do what I need to do to stay upright, keep moving, but then make a change. I do have an option here. Once at the Toe Head shore I could turn left, not right. Turning left would lead me to the surf-ridden shallow shores of Scarista Bay. Not great, but got to be better than a suicide bid round Toe Head. If I can get through the surf then a wide bay, filling with tide, will guide me into an inner loch and shelter. I will then have to portage all my kit through to the west, an option that would take hours longer than paddling round.

What to do? Before I left for this trip, my Kiwi pal Andrew, himself an experienced outdoorsman, reminded me, 'In those marginal moments, mate, trust your gut.' Right now, my gut was saying I could die out here on the wrong side of a formidable headland. So, brain frazzled and with three weeks of difficult seas and coarse weather behind me, I decide enough is enough and turn left.

Ahead lie the white breakers of Traigh Scarista with the promise of the lagoon beyond. The sea turns from dark grey to turquoise as the sand-covered bottom rises up in the shallows and, with the same shallowing, the waves rise and break all around. Somehow I get lucky and weave a way through the breaks, letting some pass, paddling ahead of others, veering off down a rising wave, glancing nervously this way and that. And then I'm through. A flush of relief and elation as the tidal lagoon, fast filling on the rising tide, stretches ahead; calm, no swell, just a stiff breeze and rain spattering the water.

Nearing the shore, I see a small bank of pink sea thrift, past its best but still lovely, carpeting the high tide mark, and the long stretch of lumpy machair that is my portage looming large. It rains then – of course it does – proper rain, hard heavy drops, so I sit as patient as the animals, waiting it out. But my kayak jacket leaks more water down over my upper back and arms, and I start to get cold. After ten minutes or so, the rain eases to a wind-driven mist so I clamber out, the Atlantic dripping from me, and the portage begins.

Not much to say about that really. Portaging is just graft. To and fro with kit and heavy boat, trying to keep stuff dry and generally failing. 'Treading the mountain down' as my favourite Scottish poet Norman MacCaig has put it, until the west side of the Toe Head peninsula appears and the wide beach of Traigh na Cleabhaig frames the Sound of Harris ahead.

I am properly tired now. I have been at it for several hours and am physically gubbed from the portage and the nervous paddle. I pitch the tent up above the beach, crawl in and sleep.

- Route: Taigh Bhuirgh to Traigh na Cleabhaig – paddle and portage
- Distance: 8 km paddle and 1 km portage
- Weather: wet morning and very windy, calmer afternoon, cloud and rain throughout
- Wind: Amber/Green. Force 5/6 dropping to Force 3 to 4
- Sea state: unpleasant. WSW swell with Force 3 to 4 blowing against it – too much for me today
- Events: swell, fear, dodging the surf relief, rain, portage
- Camping: south-west-facing out to Sound of Harris and Pabbay mist-shrouded in the west.

Hiding away at Taigh Bhuirgh

Portage at Toe Head

Windblown camp

Monday 27 June, Day 25: Weatherbound at Toe Head, 0 km (wind and wet). Wind: Amber/Red.

Weaving through the cattle to walk to the remote medieval chapel of Rubh' an Teampuill, south-east of Toe Head, I get in the mood for some history. At the end of the road, a half-hour walk from my tent, is the small but perfectly formed Seallam Centre.[6] As well as it being a warm and friendly place to recharge your phone and backup batteries, if you have an interest in pursuing your Highland genealogy then Seallam, tucked under the hill from the winds of the Sound of Harris to the west, is the place to be. The brainchild of Bill and Chris Lawson, here is where your search may track down your predecessors, drawn, as they may have been, over time and space into the global Scottish diaspora. Seallam is a place where identities are secured, where lost families are discovered, where cross-continent connections are made, and where a sense of place is formed. The Highlands of Scotland may be only a small part of a small country, but our culture, our language, our ideas, our music and our people have had an outsized effect across the globe. Highlanders have ventured far, and where they have disappeared from view, it is the research at Seallam that is bringing them back home. So I spend a couple of hours recharging batteries, yes, but also immersing myself in the exhibition and the genealogy work there.

Back down on the shore, I find the wind is already rising another notch and, with the forecast downloaded, I'm anticipating more rain and big winds again tonight. With tent poles still prone to snapping, I shift the tent to a more sheltered spot, write and read, and then walk back out to the track end once more for the highlight of my day, meeting my wonderful old friend Iain and his fishing pal Ben for dinner at their rented cottage outside Leverburgh.

Iain and I met in Lochinver some 30 years ago and have shared a love of the mountains and the wildlife therein ever since. While he is a GP by profession, his knowledge of fish, butterflies and Scottish Highland history is positively encyclopaedic, and listening to him and Ben talk fishing is a joy of discovery: remote lochans, fly types, late

---

6 It is correctly Seallam! Centre, but that looks odd here so I've removed the ! ... apologies to any grammar gurus.

risings, fighters, best catches and the ones that got away. Their love of the animal and passion for the task is matched only by their obvious love of this place. Entirely coincidentally for me, they have two weeks out here before returning to wives and families, two weeks when I fear for the safety of any wild brown trout in the vicinity.

Iain and Ben are hunters of fish, and hunting excites passions and discussion within both pro- and anti-hunting folk alike. In the UK, we hunt freshwater fish, sea fish, deer, grouse, pheasant, smaller game birds, shellfish, foxes, mink and others, and opinions quickly polarise. Within the debate there is as much feeling as there is knowing, and the space between the various parties is almost always fraught and opiniated, and often aggressive; not helpful, but perhaps to be expected. Killing or harming animals for sport is emotive – of course it is. Suffering is involved, the motivation is generally not hunger (not here in the UK, anyway), and the die-hard traditions of fancy dress (think fox hunting) can alienate as well as attract.

But not all hunting is bad. From the very opening of life on our planet, predators have played a key role in maintaining some level of species balance, albeit a naturally bouncy one, and without them ecosystems would boom and bust harder and more often. But that balance has been disrupted. And here in the Highlands, it has been happening for years right under our noses. The issue here is that those ancient predators, our wolves and bears, have themselves been culled to extinction, and the only apex predator left is us. Unlike the wolf's, our human predatory activities are generally not guided by any existential need, not any more. Instead, our hunting centres in the main around making money or pursuing some not-so-subtle social-climbing agenda. And the result of our human-centred hunting activities has been an upset of balance and a disintegration of local ecological biodiversity.

Our red deer, the so-called monarchs of the glen, sit at the heart of this issue. Our mountains – open, wild, magnificent and allegedly natural – are only open because the deer have eaten all the baby trees. The lower slopes should be alive with woods and birdsong, and where we have removed the deer, that is how they are. But beyond the fences our red deer scratch an ever more meagre living on ground sparse with nutrient, and their well-being relies utterly on us feeding them in the winter and then culling them off to keep the numbers down. The red deer numbers are artificially maintained. A more unnatural 'natural' I find hard to imagine.

And what is it that keeps our deer on the Highland hill and our slopes a relative biodiversity desert? It is hunting. Hunting with a Victorian hat on: stalking. Hunters don't come because they're hungry. They come, in the main, to shoot the big-antlered stags, to have someone else do all the work of carrying and gralloching,[7] and to live, for just a short while, the pretend life of a Victorian aristocrat in a fancy lodge. The insides of these lodges are generally a picture of privilege, and whether you stay in Sutherland or Argyll, *Country Life* magazine will grace the coffee-tables, and dead heads will decorate the walls.

Now if all this stuff was just social aspirational nonsense I'd have nothing more to say; each to their own, after all. But it isn't. The issue with grouse shooting and deer stalking is that these activities are sustaining a land ownership and a land use system that is, in my opinion, outmoded and ripe for change. Not only does keeping deer on the hill – and, indeed, burning vast tracts of moorland so that one particular type of bird, the red grouse, can grow and prosper on the new heather – limit the potential biodiversity in our landscape, but they also perpetuate a historical and arguably anti-democratic land ownership system. Our Scottish landed estates are vast holdings owned by the wealthy few – and that system, coupled with a Victorian image of what the estates should look like and how they should be run, perpetuates a poverty of biodiversity and social exclusivity that feels out of place in an increasingly social democratic Scotland.

There are a few exceptions, and I would cite the Ardtornish Estate in Lochaline as one. The Ardtornish owners, directors and management team have a progressive vision that is already making dramatic improvements to both the local ecology and social inclusion. Their enterprise is a shining beacon of what's possible if we change the way we think and then bring keen minds to bear on making changes on the ground. With a reforesting programme, affordable housing, community allotments, serviced house plots, improvements to the harbour, and

---

7 disembowelling

even the estate's own hydro scheme, this is the sort of biodiversity and socially aware estate of the future I'd love to see more of.

What 'ownership' should look like is another issue and, having introduced alternatives to the Victorian idyll in discussions around community ownership earlier in this trip, I will expand on this later on. Further south from here, within these Outer Hebrides, community initiatives are flourishing and a new future for land ownership is taking shape.

But for now, when we think about hunting and whether it is good or bad, necessary or not, there is nuance at every turn.

Not all hunters are bad, of course. I know hunters who not only have deep knowledge of the ecosystem they inhabit but who also care deeply about the animals they kill. Some are still bedded into traditional ways of life on the Victorian estate, but others, like my friends Iain and Ben, are fully modern in their attitudes and acutely aware of the wider issues surrounding their activity. If we are to judge this activity – and plenty of anti-hunting people do – then I think understanding 'intent' is key.

Intent is visible through the lens of each of our knowledge bases and/or what we think a given landscape should look like. Hunting for some will be just a traditional country pursuit devoid of intellectual progression and science, while for others it will be intimately connected with efforts at genuine conservation and biodiversity renewal. As I explored amongst the rhododendron forest back in Argyll three weeks ago, culling one species to protect another is part of that conservation debate, and in New Zealand and Australia, as Emma Marris points out, that debate is raging. Here in Scotland, where we appear to do the opposite by preferencing one or two destructive species over the many, we certainly have cause to hold more robust discussion on the subject. But conversations seem few and far between. We have culled a few rats and hedgehogs to protect at-risk bird populations on islands, and we routinely cull foxes, badgers, crows and pigeons to protect our plant and animal food crops. But all that culling is hunting with purpose – to protect food, to recover species. Hunting for pleasure, however, seems different.

Hunting for pleasure, for 'sport', feels like it inhabits a rather different moral ground. Here, for most, I suspect the intent is to have fun, perhaps to pursue some social agenda, to compete with others, or maybe to demonstrate skill with a gun or rod. But pursuing such sport brings certain species of animals into the world *en masse* – think pheasants or grouse, for instance – purely in order to kill them. Yes, the killing produces a little food, but that's not by design, and it doesn't have anything to do with protection or with creating a more biodiverse ecosystem. Indeed, one could argue that it encourages the opposite, and the fierce penalties around illegally killing raptors here in the Highlands provide a window into what would happen were sport shooting allowed to run rampant.

And what about fishing? Does fishing for fun have the same moral stickiness as that for birds or deer? Until recently it used to be thought that fish did not feel pain. We know now that that's not true. Fish have pain receptors in their brains – and that, coupled with the obvious stress they demonstrate on the end of a hook and line, and as they're hauled out of the water by whatever means, cries out, or at least should do, for our compassion.

Full disclosure here. I fish, or at least, I have fished. I've hunted trout with flies on remote Highland lochs, and I have drawn up denizens of the deep from the sea. The latter have always been for food, and once you've caught a mackerel or two you can stop fishing; your fridge is full. But the former, the freshwater fishing, I have pursued for a different reason, and it's not because I'm hungry. Laying a line out on a remote Highland lochan is addictive; there is something about being out in the open country, the mountains and lochs as companions, the weather coming about you, the quiet, and of course the lure of the unseen adversary beneath the peaty water. There is something in all that that satisfies. Something around the tonic of being outside, the wide-open space, the excitement of the chase. Sometimes I catch and sometimes I don't (actually, mostly I don't; I'm not that good at it), but by engaging in the hunt in that place, there is something in there that draws.

And in the hunting, I'm learning. Learning about the animal and its ecosystem, how it behaves in different weathers, at different seasons, what it eats and why, and whether it's thriving or suffering. And for me, if it stops at that, then the activity is indeed devoid of value. But if I can take that knowledge and use it to engage with the animal's wider ecology, then perhaps I can justify my hunting pleasure. To advocate for change, to contribute to projects that increase species' chances to thrive, to increase ecosystem biodiversity – *that* is where the value is in fishing. As a fisherman, I should be self-aware enough to recognise that if I'm going to hurt the few, then I have a moral duty to benefit the many.

It is all about intent.

/ / / / /

My head buzzing with thoughts, with the wonderful conversation at dinner, my hosts' conviviality and my belly full of hot food, I stroll back to tent at 11pm with the gale still blowing about me. And it's still light.

- · Route: in situ at Traigh na Cleabhaig
- · Distance: 0 km (weatherbound)
- · Weather: wet, wind and more wind
- · Wind: Amber/Red. Force 5 to 7 south-south-east
- · Events: gannets diving offshore, chapel, Seallam genealogy, wet, windy and noisy night, dinner with Iain and Ben
- · Camping: sheep-shit-ridden spot in the grassy dunes, but just enough shelter from the near-gale.

Tuesday 28 June, Day 26: Weatherbound at Toe Head, 0 km. Wind: Amber/Red

From my wind-blown camp above Traigh na Cleabhaig, I read, write, walk between showers, climb Toe Head peak, Ceapabhal, in the driving wet, yoga session on wind-strewn beach, sleep and eat. Forecast is suggesting I might get off tomorrow afternoon. Thursday is still forecast for lighter winds, too, so I'm feeling hopeful. I'm now in the right place to take advantage of the wind dropping, and the risks taken east of Toe Head and the graft of the portage should now pay off.

- · Route: in situ at Traigh na Cleabhaig
- · Distance: 0 km
- · Weather: wet, wet, wet … relentless, wearing, everything is damp

- · Wind: Amber/Red. Force 5 to 7 south-south-east
- · Sea state: it's just white out there
- · Events: walk up Ceapabhal, rain relentless and wind blast-
  ing through the jacket; all my kit is at saturation point
- · Camping: same as last night.

Wednesday 29 June, Day 27: Toe Head to Lochmaddy (head of the loch), 27 km plus portage. Wind: Green/ Amber.

Today I break free of the surf zone at last! The day has dawned cloudy, showery and windy as ever, but forecast is for a slight easing, down to Force 3 to 5 around midday. Looking out to sea I can see no change, but watching the surf dump on the beach suggests that something's happening. It might just be the tide, of course, but at around midday it looks kind of okay. As ever, all is marginal and, with wind in the face, I think I might at least haul my way south-east towards Leverburgh and make another decision there. There is also the tidal stream to consider. Ahead lies the Sound of Harris, a place of strong tidal flows, a complex maze of skerries at low tide, and open water at high tide when the skerries disappear under water. I have come across the sound before and I remember that the exposed skerries can provide some shelter and occasional respite from the wind. With wind from the south-east, low tide due around 2pm and rain showers becoming fewer and further in between, at lunchtime I decide to go.

Boat and kit ready, and then I'm standing once again at the water's edge, counting waves and trying to judge the launch to avoid a dumping. I sense the lull and leap in, and the surge of the retreating water promptly turns me sideways to the waves. I focus on pointing my bow back into the surf and thus do not get my spraydeck on before a big old breaker rears up and breaks onto my chest, giving me a mouthful of sea water in the process. A couple of strong strokes and I'm through.

Winded from the weight of the breaking wave on my chest and spluttering from swallowing too much salt water, I manage to get the spraydeck on in the bouncing sea outside the surf line, and I'm off! But the boat feels sluggish and slow, and I only realise when stopping an hour or two later that I have several litres of water sloshing about around my feet. I never noticed simply because I've been focused on paddling the lumpy sea on the nose, and there's no way I'd have stopped to pump it out anyway, not in that weather.

I haul up to the tiny island of Sromaigh and rest for a minute under its sheltering cliff with the wind rushing by on either side, taking time to assess the next move. Looking, looking, looking. Checking the sea state and tidal movements. I want to cross to Ensay, just a couple of kilometres south-west, but the tide's running hard past the light-rock ahead and the fast-emerging skerries round about surf with swell. I take a deep breath and go for it. Into crossing mode, up the pace but keep the form, push through the tidal mess and grind it out against the wind to gain some respite off Ensay's north shore. The distances are small but in this wind they sure don't feel that way.

After that the plan works well. The wind remains fresh all day and at one point begins gusting strongly too. But I am indeed able to use the exposed skerries to leapfrog across the Sound of Harris at last. In the middle, it feels exposed; I'm a long way from anywhere and surrounded by sea and rocky islands. It's gorgeous, though, and I see several otters, including a mother and baby, playing close to the boat. So close in fact that I can see their whiskers twitching and the individual hairs on their coats. Seals too come visiting and moan at each other across the empty spaces. Terns, dipping and hovering above, screech from their colonised islets, and gannets and gangs of razorbills, fast, focused and full of intent, fly low between their fishing grounds in the Minch and their western island summer homes.

And then there is the light. Exquisite, fresh, clear, sparkling light wherever sea and sky compete for effect. Vast cumulonimbi are forming in the south-west, their tops bright white in the sun and their bases a forbidding grey. Around me, as I cross between skerries, the light changes from grey to silver to rainbow to a touch of sun, and the sea changes colour and feel with it. A grey sea and sky with wind can feel threatening and a bit forbidding somehow, but if the sun comes that same sea lifts to become something all a-sparkle and just a fine place to be.

On the far side of the sound, my body now tired and legs and back aching, I push ashore on the northern tip of Taghaig, and the sun comes out in full. A few very ancient black houses with rounded walls become my drying lines, and I take the chance to eat and to stretch; these last two weeks it has been so rare that I've had a chance to relax outside without looking over my shoulder for the next rain squall.

The scores of islets and inlets leading to the portage into Lochmaddy itself lie ahead. Map-reading is very tricky in that maze, and while I do get to the place I've aimed for, it's not the best spot to portage the gap, due to the tide still being too low. This means a long old carry over rough ground. There's another two hours before the high tide will allow me to paddle in close to the land bridge, and a more patient man would have just waited. I am clearly not that man so I set off loaded with kit, including my tent, to walk over the gap, currently almost a kilometre wide and over very rough ground. Will I ever learn? With rain now imminent, I pitch my tent just above the tidal zone on the Lochmaddy side. On returning to the boat I spot another inlet much closer to the road, and decide to paddle the boat and remaining kit round to this second inlet. Back on the water, I miss the alternative opening completely in the maze and manage to add an extra kilometre to the portage. 'It's just graft' is the comforting refrain, but when it's self-inflicted and unnecessary, it's not so comforting. Two hours later, in the rain of course, I have everything in one place and a brew on. Gubbed.

Waking in the early hours before dawn, what I notice most is the silence. After days of booming surf, the quiet of the night is just wonderful. To begin with, as sleep slowly fades, it takes a few moments to register the quiet. You hear nothing – utter stillness. Then come the tiny sounds, some near, some far. But these seem peripheral to the silence, and somehow add to the quality of it. Quiet – a distant gull – quiet – a sandpiper's peep – a flutter of breeze – quiet – nothing – silence – sleep.

- Route: Traigh na Cleabhaig to head of Lochmaddy (including portage)
- Distance: 27 km
- Weather: early wind and grey give way to sun, then rain later
- Wind: Green/Amber. Force 3 to 5 on the nose to Ensay then drops to a 2 to 3 for the rest of the day
- Sea state: lumpy start on the nose, then skerries take away the swell and choss, and just a windy sea after that

- Events: punching through the surf, tidal flows in the sound, skerry-hopping, seabirds' corridor, lunch stop at the black houses with sun out, the portage errors, the quiet of the night
- Camping: just above the high tide mark at head of Lochmaddy.

Thursday 30 June, Day 28: Head of Lochmaddy to mouth of the same, south-east side, 11 km. Wind: Green/Amber.

Leisurely rise, the first for a long time, as have to wait for high tide to allow passage south through Lochmaddy's islands and skerries. Lochmaddy is the anglicised version of Loch nam Madadh; the Gaelic *madadh* means 'wolf' or 'hound', which in this loch system relates to the many rocky outcrops and islands dotted throughout the exposed outer bay. I like to think of them being named by fisherfolk of long ago in boats driven by sail and oar. Returning to the loch from the Minch after dark, boats laden with fish and with wind whistling about, these rocky *madadh* lie in wait, ready to finish off the unwary boatman in a splintering of wood on rock.

Paddling south-east now, I find there is more wind than forecast but I am at least in sheltered waters. Wind on the nose, outgoing tide, several flows, small overfalls in parts, so I'm fairly speeding along. Geese, otters, herons, one large raptor, terns, occasional red deer, great northern divers, mute swans, and a gorgeous sighting of a small flock of shelduck arcing up off the sea across a big sky with the seaward hills as backdrop.

It is cold today, and by the time I put in at Lochmaddy village I need warming up. Change into dryish clothes while tucked behind a rock at the eastern slip, then walk up to the shop for the planned resupply, and off to the hotel for a feed and a recharging of phone. No mobile signal in the village, so take the opportunity of wifi to touch base with Leah before heading out into the open country again.

Paddle off into rain and a stiff breeze from the south-east, and several kilometres later I am camped right on the north-east tip of the rocky peninsula at the south-eastern end of the main Lochmaddy Bay. Although I know a sea will be running, I paddle round the corner to have a look at what tomorrow's journey might bring. Heading south from here will involve an exposed section of coastline, and with today's blow the sea's a mess. The wind is whistling through the gap between the mainland cliff and the Madadh stacks just offshore, so I sit in the waves for a couple of minutes, cogitating. If I went any further I'd be totally exposed, and if it all went wrong it's all cliffs and jagged rocks, no get-outs for 6 kilometres, and it's unlikely anyone would see a flare out here. I opt for discretion not valour, and turn tail to a tiny bay just shy of the point. In stark contrast to the busy sea round the corner, it's very quiet here and somehow slightly edge-of-the-worldly in this grey weather. The ground beneath my tent is sodden, so hopefully I'll get away tomorrow, when the wind is supposed to drop.

Ahead lies some remote country, several days alone, no mobile signal (bliss!), so I'll need to go canny. I have time on my side, and I still don't have to be anywhere by any specific time. That is the huge difference on this trip; the relatively open end to the journey means I can give myself time to make decent decisions, I hope, and – more importantly, it turns out – allow myself to manage the somewhat extreme so-called summer weather.

- Route: head of Lochmaddy to south-east tip of Lochmaddy Bay
- Distance: 11 km
- Weather: mostly cloud and rain, bits of sun in morning
- Wind: Green/Amber. Easterly and south-easterly, Force 3 to 5, cold
- Sea state: wind-blown only; nothing big, but always on the nose
- Events: poking nose round the corner at the mouth of Lochmaddy with a big old sea running
- Camping: feels remote but am well poised to head south. Strong sense of the big sea just round the corner. Quiet, green, wet, hilly, heather and grass – quite different from the west side of Harris.

5 | Living with uncertainty

Days 29 to 39 – Lochmaddy to Castlebay (Barra) to Oban

Friday 1 July, Day 29: Mouth of Lochmaddy to Flodaigh Mòr, 16 km. Wind: Green.

Wet night, calm night, calm morning, condensation clinging to the inside of the flysheet, strange dreams, pee breaks galore in the wee small hours, slight sense of intimidation from the sea running high just round the corner.

Clamber out of the wet tent at 5.45am and splash out across the bog to the headland overlooking the Stac of Madadh Mòr. The wind is off. Yes! Sea state right down. *It's a goer!* By 7am I'm paddling off the rocky beach and round the corner, heading south towards Loch Euphort and beyond.

As I make the turn south, tiny as I feel in the kayak, there is yet again that excitement of a journey progressing and a new horizon appearing. The sea opens up vast and expansive to my left and the cliffs loom huge and dark to my right. Some of the swell from yesterday is still in the water and is bouncing back from the cliffs. I am a wee bit nervous of the potential for the wind to return and leave me exposed here on this rocky shore, but with the sea as it is I feel more relief and excitement than angst for sure.

And then the sun comes: grey sea to blue, dour hills to sparkling green after the wet, and a big sky complex with clouds I struggle to name. A sea eagle soars out from the cliffs, mobbed by gulls, terns

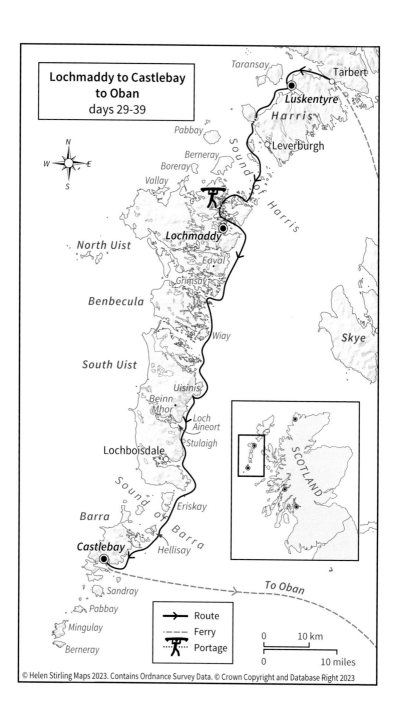

Lochmaddy to Castlebay
to Oban
days 29-39

N W E S

Taransay
Tarbert
Luskentyre
Harris
Pabbay
Leverburgh
Berneray
Boreray
Vallay
Sound of Harris
Lochmaddy
North Uist
Eaval
Grimsay
Benbecula
Wiay
South Uist
Skye
Uisinis
Beinn Mhor
Loch Aineort
Stulaigh
Lochboisdale
SCOTLAND
Sound of Barra
Eriskay
Barra
Hellisay
Castlebay
Sandray
Pabbay
Mingulay
Berneray
To Oban

Route
Ferry
Portage

0   10 km
0   10 miles

© Helen Stirling Maps 2023. Contains Ordnance Survey Data. © Crown Copyright and Database Right 2023

bob above, wings flicking, checking me out, a few black guillemots huddle in groups, their high-pitched peeping lending some tune to the more raucous squawk of the gulls. It is rough bounds here south of Lochmaddy, some sizeable hills, no houses, no people, no paths – just cliffs, steep rocky shores and the restless sea sloshing at its edge.

The mouth of Loch Euphort appears, and I decide to keep going. The sea is manageable, the sun is up, the body feels tired but okay, and it's just great to be making progress. I relax. The boat rides the waves, my body is at one with paddle and rudder, and the gentle swell and occasional flurry of breeze adds frisson to the day. It's a beautiful morning.

I want to climb Eaval if the weather allows so, observing the cloud base slowly rising, I pull into Bàgh Mòraig, just shy of the island of Ronay. Exquisite little loch with a narrow opening through which a hidden world opens up. Top of the tide takes me in, no bother, and I find a spot to pull up on a seaweed-covered shore. Kit change, boat secure and off up Eaval, the highest point on North Uist. There is evidence here of human settlement, but with the bracken shrouding the old fertile bits of land I don't see any of it. Mist is coming and going off the top, vistas of loch and land and sea open up, and a cool breeze whips the sweat away as I climb. It's lovely to be walking again. At the top I sit tucked under the summit rocks out of the breeze, and gaze out over this unique landscape that is more water than land.

Perched precariously on the edge of the Atlantic, this island is barely that. Its western side, bearing the full brunt of the Atlantic, a full 2,000-mile fetch of ocean which pounds the coastline to a line of white shell sand stretching 100 kilometres from North Uist south to Barra. Above the beaches the naturally acidic soil is softened by blown sand, which is mixed and fertilised by the owners of thousands of heavy cloven feet. The resulting grazing is a unique ecosystem known as the Highland machair, meadows of grasses and flowers that paint these marginal lands in a springtime flourish of colour. Here too, the Western Isles crofters have for centuries grafted a living, and while the tools of the trade – the tractors and the ploughs – have removed some of the back-breaking work, the essence of the crofting task has changed little. Cattle, sheep, pigs and chickens; some cereal and root crops, which provide both food and fodder, and the movement of livestock between lowland crofts and higher summer pastures, where available, have gone to create the pattern of vegetation I see laid out below.

Crofting nowadays is much more than livestock farming, as crofters must note changing markets for goods and services as much as any other business. But critically for these islands, crofting does hold its older values close. It holds people to the land, offers relative security of tenure, creates local communities, and maintains a knowledge base and a way of living that is the antithesis of big agri. Small scale, community based, environmentally aware and protected by law, the modern-day croft is slowly evolving, pivoting with the times. In a country so cowed by past feudal land ownership – much of it our own making of course – the crofters' hard-won rights and freedoms remain an important part of Highland culture. It is rather wonderful now to see those freedoms, that ancient desire for autonomy and agency, expanding with community land ownership to include whole estates.

Looking out over this stunning landscape, rich as it is with human use and meaning, I think of the pride my wife and I felt all those years ago when we too had access to a piece of crofting ground. As explored earlier in this book, we struggled financially and had to move on in the end, but we did so only after watching the Assynt Crofters make history by taking that leap from crofters to estate owners back in 1992. The excitement and sense of history-making was deeply felt by all involved, and seeing as I'm planning to pass through Assynt on this journey, I'll explore the detail around community land ownership then.

/ / / / /

The cloud base is coming and going, blurring the views of the countless lochans and scattered crofting townships below. Time, I think, to head back down to the boat.

Several hours on and I'm camped on the aptly named island of Flodaigh Mòr, the flooding narrows, just north-east of Ronay. Views back at Eaval across the water, east out to the Minch through the tidal flush, the *flodaigh*, and inland to the island-studded complex of lochs and channels that lead eventually to the west coast.

And it's *dry*! Full body wash in the loch and a yoga session on the grassy sward by the tent. What a tonic! I could have paddled on, but then I'd be missing this quiet place. To be still on a lovely day is a rarity to be savoured, especially on this trip it seems. There is no sun now, but a gentle breeze is keeping the midges at bay. I allow the landscape to settle around me, the animals and birds to visit, the light to change, the wind to blow, to die, to blow again. And of course the sea, always the sea, moving, flowing with the tide, calming, rising with the wind, falling again. There are small fish rising in the shallows, and the terns plunge-dive when they spot them; a lone Arctic skua chases a gull at speed, and a curious seal spy-hops ever closer, investigating my yoga antics.

Latterly a small two-man fishing boat appears, to lift and reset the creels lying in the channel where the tide is running. As the boat comes close I finish my yoga and squat down on the rocks above where the men are fishing. They head over and pull in for a chat. Young men, hardy-looking and obviously utterly at home on the sea, Willy and Ewan are wrasse fishermen, capturing what the salmon farmers call 'cleaner fish' for the industry. We discuss wrasse fishing, salmon farming, ethics, management, fish politics, community jobs, community funding and more. So much in common, so much passion and a determination to do the right thing. Willy is interested in pursuing the seaweed farming piece, too, so I point him to our sister company, New Wave, who are already farming seaweed in the seas south of Oban.

I love this connection with folk here, the common interests, the sense of place.

- Route: Lochmaddy (south-east) to Flodaigh Mòr
- Distance: 16 km
- Weather: cloud, some sun, light breeze, no rain until later
- Wind: Green. Light, Force 2
- Sea state: residual swell under the cliffs, then calm in around the island
- Events: turning the corner at Madadh Mòr, climbing Eaval, washing in the loch, meeting wrasse fishermen
- Campsite: exquisite, overlooking Eaval.

Saturday 2 July, Day 30: Flodaigh Mòr to Uisinis, 35 km. Wind: Green.

Wet night, almost constant rain spattering on the tent only inches from my face, noisy enough to keep me awake. Felt like a long night, too, but as I peer out into the early morning I'm reminded what a perfect camp spot this is. The rain eases around 7am and there is calm. The forecast shows the wind rising to Force 4 then staying high for a few days, and rain to come later today. I am chasing windows of opportunity again. Time to go.

The east side of North Uist and Benbecula is gorgeous. Dotted with islands, green with new grass and bracken, alive with birds, it always has another corner to turn, another view. With the tide dropping too, the tangle of different seaweed species appears and all that intertidal zone adds interest. Today I see two otters, one appearing right next to the boat and unaware of me for long enough for me to get an unusually close look. It's clearly thinking of taking a breather or something, because they are normally frenetic and it's rare to see them still, especially in the water. It sees me, dives, and performs the usual otter vanishing trick. How that happens is beyond me. They must either know the ground well or suss out which direction I am travelling in. Surely it will reappear close by. I'll wait; it's moving that way and so am I … But most times, you'll either not see them again or perhaps catch a fleeting glimpse of a flicking tail or a back-turned head before – gone.

The deer perform the same magic trick on the hill. Whichever direction you approach from, they seem to know the quickest route to disappear from view. There is rarely any pause or indecision. Heads come up from grazing as you appear, then they turn and high-trot or canter round a bluff, behind a rocky outcrop, round a contour – vanished. It's all in the knowing of the landscape. Instinctive, prey-animal, survivalist 'knowing'. Deer and otters have both been prey for humans in the past, and the former remain so of course. Those deer slow to react to an unusual figure on the hill or an unexpected movement in the heather are less likely to survive the bullet. Otters are not hunted now, but there is clearly a memory, a meme, bedded into their behaviour that reminds them that humans, generally, are trouble.

Other animals, birds in particular, offer different reactions. Geese at this time of year are protecting their flightless young. At sight of a nearby kayaker they rush away in a panic of flailing half-grown wings and spray from frantically flapping feet. Shelduck are quieter carers. They shepherd their young into the brown seaweed and one of the pair entices you away with a fake broken wing, a performance worthy of any stage show. Ringed plovers do the same. How does this behaviour evolve, I wonder? To come close, to face up your enemy, is one thing, but to add the theatrics, that is a whole other level of deceit – and all born from the desire to protect. They are in effect potentially sacrificing themselves for their young. The stress must be huge, and the more I think about these things the more I try to mitigate my own effect on their lives. I am there, I am moving, I rarely see the birds before I round a corner, and there they are. So I focus on passing through as quickly and quietly as possible, and as I do, I see the parent return to the family group behind me.

Other birds are more forthright. The oystercatcher's high-pitched and piercingly loud piped alarm call is full of aggression. These beautiful black and white birds, topped and tailed with orange-red bills and feet, are as aggressive a parent as any tern or great skua. They will pursue you, dive at your head and harry you until you get the hell out of their space. Gulls, unexpectedly perhaps, seem less aggressive. They rise and circle about, squawking, but there is no diving, no attack. Terns and bonxies, however, are another thing entirely. Watch your head; blood may be drawn.

Today, paddling along between islands and across the mouths of lochs, I am accompanied by all these birds and more: mergansers, eiders, black guillemots, a sea eagle, ravens, redshank, sand pipers, ringed plovers and all the gulls. Another treat is the lone red deer feeding on a small low-lying island as I pass. It sees me coming, runs along the skyline at pace down to the shore, plunges into the sea and swims, head high out of the water, eyes bulging, to apparent safety on the main island group. I have clearly ruined its day.

Throughout this journey I've been blessed by countless wildlife encounters. I have tried to observe with an open mind and to be aware as to what impact I – indeed we all – might be having on these animals' lives. The wider impact of climate change on species is well enough understood, but from my perspective at sea level I don't see species. Instead, I see individuals, and I wonder if I can observe these individual souls to better understand my own low-level impact on their behaviours. All animals, including ourselves, act on instinct, and this is especially obvious when we are fearful or hungry or horny. But many animals are also making conscious decisions too. Most of them, when carefully observed, can be seen making decisions, changing their behaviour depending on what's happening around them. Even fish, long thought to be without conscious thought and, as mentioned earlier, even not to feel pain, do both. Wrasse, groupers, eels and many other species not only cooperate but communicate – and yes, they do feel pain. In the wild this is difficult to study and takes many hours of painstaking work. However, in the more controllable farmed environment, much can be learned, and perhaps, then, we can make some informed guesses on the lives of those animals beyond.

When trying to understand farmed livestock, you need to start from a position that recognises that the animals are farmed and not wild, and that they are unlikely to exhibit exactly the same behaviours as their wild cousins. That fact is too much for some people, and an anti-farming movement is active worldwide encouraging us all to see farming of animals for meat as cruel. I agree and disagree with that sentiment: yes, performed badly, farming can be cruel – but when it is done well, as envisioned by a number of animal welfare NGOs here in the UK for instance, farmed animals can, I believe, live a good life. The trick of course is understanding this as objectively as possible.

In the wild, countless millions of fish, seabirds, land birds, mammals, reptiles and insects die unpleasant deaths. Disease, weather and predation take the lion's share. Old age and a peaceful death are, I suspect, a luxury for few. Most of their deaths are invisible to us, as sick or frail animals hide themselves away, and predatory or scavenging animals profit from the energy transfer in the eating. It is the natural order of things, and species boom and bust in an endless cycle of scarcity and plenty. Are these wild animals living a good life? Well, they're living a wild life for sure, and they have agency – but for many, as for early humans perhaps, their lives are likely more Hobbesian:

North Uist from Eaval, more water than land

Salmon farming off South Uist

East of Benbecula
and wet…again!

The remote Uisinis bothy on South Uist

'nasty, brutish and short'. But even that is an anthropomorphism, and maybe we should accept that we cannot really know what it's like to be another species.

Or can we?

Having been involved with the Scottish salmon industry over some years, I have spent much quality time engaging with farmers. I was a farmer myself for a while, too, and have worked throughout the food chain, gaining insight not only into farming but also processing, retailers, NGOs, anti-farming lobbyists, academia, regulators and government ministry. In my kayaking journey this year I have passed many salmon farms, all of which I know in some detail, having visited and assessed them over the past few years on a range of fish health, welfare, environmental, social and economic criteria.

What has all this got to do with understanding what the wild animals I have been seeing on my journey are thinking as they go about their daily lives? The answer is that over the last few years, together with a leading UK retailer and a range of experts in the field of animal welfare, we have been taking a deep dive into attempting an objective understanding of the behaviour not just of fish but of all the farmed animals under our responsibility. The reason for the work centres on a desire to improve both animal welfare and farmers' knowledge of the animals under their care. Findings thus far indicate there may be useful themes in here to help us observe and perhaps recognise similar behaviours in the wild.

The process of understanding is complex, and starts by us asking ourselves how can we know whether our animals are living a 'good' life? Sometimes folk insert the word 'happy' here, i.e. are our animals happy? I am sceptical of this word, as it feels assumptive and is too anthropomorphic. Instead, we need to attempt to understand the animal on its own terms, within its own world, and find words to describe behaviours that help us understand whether our fish may be living well. There is lots of good science surrounding this work. At its heart is a process known as 'qualitative behavioural assessment'. This is a system of recording observed behaviour on a regular basis, in some cases several times a day. Descriptive words are decided upon by teams of farmers, vets, biologists, academics, NGO members and others, and these terms are then split into four categories, from high to low stress, and from active to passive behaviours. Counterintuitively, perhaps, some passive behaviours can be negative and some high-stress ones positive; the upshot depends on the species and their particular environment.

The farmed environment for all livestock species is not static. There is feeding, cleaning, milking, mating, gathering, treating, reproducing, sheltering, resting, socialising and more. Some of these are imposed activities which are part of the farming process; others are arguably more natural. But within these environments our animals do exhibit observable behaviours. To understand what we are seeing, however, needs more than a cursory glance at the pen of fish or the field of cattle. Real observation needs work, and it needs data. The latter is gathered over time, and patterns of stress or the lack of it are slowly revealed. Where learning occurs, farming techniques and behaviours are then changed to create a pattern of ongoing improvement based on a growing knowledge of observed behaviour.

So my opinion, from my own experience of farmed animals, is that it can indeed be possible to better understand whether animals are living a good life. And in the farmed environment, if we want our animals to thrive, to live with minimal stress and to grow well then understanding whether they are living a good life must become a given. It's not a perfect process, and just as we can't assume what any of our fellow humans are feeling at any point in time so must we retain a healthy scepticism that we can do so for other fauna. But we can try.

Beyond the farm gates we don't have the same control or observational data for the mass of animal life out there. There are exceptions, and there is plenty of evidence-based work performed with apes, big cats and other mammals that captures in detail lives being lived in the wild. With birds or fish, however, it is a little trickier; we can count numbers of animals between breeding seasons and we can observe individuals as best we can, but we have to accept that most lives are lived beyond our horizon and that many may indeed be 'nasty brutish and short'

/ / / / /

The rain has come and gone and come again. My back and shoulders are wet once more, and I take heart that I can at least get dry, or at least drier, in the tent at night. Meantime, toughen up!

Ronay falls astern, Eaval starts to lose its relief with distance, the Maragaidh Isles loom then fade, and the salmon farms I know well around Maragaidh and Greanamul appear out of the rain, then they too fall away behind. I'm making good progress. Original plans to explore Loch Uisgebhagh are abandoned – today's wet and cold breeze is not conducive to wandering, and in these conditions standing on a shore getting cold would be no fun. So I paddle on.

I am kayaking within the Uists. Two big islands, North and South, with Benbecula, the island that is barely an island, in the middle. Indeed, as mentioned earlier back on Eaval, all three of these jewels in the Atlantic are barely islands. Look at a detailed map and you will see that Benbecula and North Uist in particular are mostly water, only just rising above sea level. South Uist is different: low-lying like the others on the west, but rising to the heights of Beinn Mhòr and Hecla on the east, exactly where I'm headed. For a kayaker these islands are a kind of paradise – in good weather! In coarse weather they're still wonderful, just more challenging. Wide-open skies and windswept seascapes in which you can lose yourself amongst myriad islets and skerries, and stumble across otters and birdlife almost every time you turn a corner. Named by Norse or Gaelic speakers, depending on who you read, and referring simply to 'west' or 'dwelling', their names have another possible Gaelic meaning: the 'crossings' or the 'fording isles'. Etymology apart, these latter names fit my imaginings of ancient folk travelling by foot or by small boat, crossing the tidal flats and flows between the island groups and capturing their essence, their physical reality, in the naming as they went.

These Uists are all these things and more. They are places of dwelling and working, of people with a keen sense of place, of community-owned land, of fisherfolk, of entrepreneurs and of course of extraordinary natural beauty. The sea lochs of Eport, Sgiopoirt and Aineort probe deep inland towards the western Atlantic, and at the vast tidal flats straddling Benbecula the eastern and western seas do indeed meet in the middle. Here, the play of sun, wind, water, skerries, rocky shores and wide-open skies do something to my head. I am utterly captured by this place, it seems. Every time I visit, for work or play, it just fills me up. Hard to explain, easy to feel. It is simply joy, I think. Joy that such physical beauty exists, that people and wildlife here seem to live more harmoniously than elsewhere – and that the wild animals have space and agency of their own. Mother Nature rules here, and I am, for a few short weeks at least, utterly free amongst it all.

Paddling down the east coast of Benbecula, then taking the outside passage around the east side of Wiay Isle, I turn west against a stiffening north-westerly breeze to grind up between Steiseigh and Lingeigh: beautiful twin islands of green, shores painted white and yellow with rock lichens and sea pinks, and always the glistening russet browns of seaweeds at low tide. There is wonder in the tangle, too, with a wide variety of plant and animals making a range of livings from the ever-changing environment in the intertidal zone. Then it's a left turn across the wide-open Bay of the Gulls, Bàgh nam Faoileann, which opens westward to the Atlantic and is effectively the southern sea border of Benbecula. South Uist lies ahead.

I consider camping in one of the bays in Loch Sgiopoirt, but the lure of the big hills on South Uist is too much. There is the wind too, of course; forecast to build from this evening with Force 5 to 7, so there is a strong chance that in the loch I could get stuck on shore for a while. A better option would be to get round Uisinis Point and take shelter at what I think may be the remotest bothy in Scotland, up above Uisinis Bay, about 10 kilometres away. The breeze is freshening, and my back and legs are sore from sitting too long in the boat, but down at Uisinis at least I'd be safe.

And what a magnificent 10 kilometres it is: the mountains of Hecla and Beinn Mhòr looming out of the sea straight ahead, and Uisinis Lighthouse perched precariously on the point, at this distance just a tiny dot but marking my target for the day.

Some shoogly tide and wind at the point see me around and under the cliffs, and it's a very tired paddler who hits the rocky Uisinis Beach. At low tide, of course …

The bothy is well up off the shore, and the beach itself, especially at low tide, takes some effort to get boat and kit up safe. The day is not done till it's done, though, and I am blessed with a dry evening to do the four trips up the hill to the bothy. Now I'm safe inside, wet kit hung up to dry all around me, my belly full of tuna pasta and copious teas on the boil.

- · Route: Flodaigh Mòr to Uisinis Bay
- · Distance: 35 km
- · Weather: calm, cold, stronger breezes later, and lots of rain coming and going
- · Wind: Green. Calm early, breezy later, Force 3
- · Sea state: generally calm but some windy wet crossings. Bàgh nam Faoileann and the crossing down to Uisinis Lighthouse, lumpy with tide and wind
- · Events: seals playing off Flodaigh, otter right next to the boat, deer swimming, rain showers, islands in the sea, contrast between the North Uist hills and serious-looking South Uist country ahead, lumpy tide at Uisinis point, tough carry up Uisinis beach and to the bothy
- · Camping: bothy.

Sunday 3 July, Day 31: Weatherbound at Uisinis, 0 km. Wind: Red.

As I peer outside the bothy at 4am I see an owl quartering the ground below me and just above the beach. Silent wings, silent hunter.

A scuttling above my head, a rushing about on tiny claws, a scratching. Mice, or something bigger, in the rafters. It is early, too early to check the time, the day barely begun, so I huddle back deep into my sleeping bag, and sleep.

As night turns to day the wrens put forth. A more convoluted, gregarious sound I cannot imagine, and from a bird so small. They're nesting in the lichen-covered walls above the bothy, walls slowly being recovered by nature, ecosystems of their own, full of life.

Later, on checking the boat, I listent to two loons, great northerns I think, sounding their characteristic chuckle out in the bay. Private birds, wild birds, the sound of places far from human habitation.

Otters come visiting, not doing much, just passing through, already fed perhaps.

A splash in the bay. I turn from climbing back up to the bothy, and two seals are frolicking. Spare time from fishing, or maybe a mother and young. I've seen seals fighting before, and it isn't that. No, 'frolicking' will do nicely.

Wind and more wind. Hustling the bay and me all day. Westerly gusts piling down from the big hills behind. First the smell of rain and then the rain itself. And what rain! First it spits, the gentlest of warnings, and then opens to full bore, a soaking in seconds. Sheets of water blown sideways recolour the landscape a uniform grey, a moving grey wall, same colour as the sea. The drips from the roof turn to streamlets, and the drops splash into the plastic drum at the bothy's edge. The birds go quiet. How they survive these onslaughts I do not know. Perhaps they don't. Perhaps out there, hidden from our view, this seeming endless summer of wind and rain is killing off the small and the young. The goslings, fluffy with down, the small birds such as the wren, the tits, the pipits and more – how do they survive? Perhaps, like blue tits in the winter, they huddle together in larger broods, the Jenny wrens tucked away deep in the walls, surrounded by a bed of moss. But what of the ringed plover, the sandpiper, the golden plover I saw high up on Eaval? These birds nest on the ground in tiny hollows, exposed to the elements. I feel for them all.

Deer, red deer, all hinds, startle as I appear out of the bothy between showers. Down for the grazing around this slightly drier ground, no doubt, and not used to seeing humans about. They're looking well kempt on the extensive grazing in these remote glens. One of them barks. I don't see her at first – she's above me and behind the walls, but she's close. She barks once, twice and again, and then I see who she's signalling: two large hinds on the boulder beach 300 metres away, grazing in amongst the seaweed and fully in my view. They raise their heads at the third bark – maybe that's the time it takes to hear things above the breeze – and there is a clatter of loose stones and a gorgeous, apparently effortless, high-footed trot over the boggy heather-strewn ground over which I splash with such effort. And they're away.

Later this evening, a rainbow. The full arch, doubled up. I have no sun above me, but there is a momentary brightening, an opening in the clouds out over Uisinis Point. Vivid colours set against the wall of grey rain either side. A splash of loveliness on a cold wet day.

All day the wind has blown hard, rising to Force 6 mid-afternoon and turning the sea white. I've had a day doing stuff: stuff for me, stuff for the bothy. There is now a new drying line ready to go – easy to dismantle, but obvious, I hope, for others to set up in seconds. There is more wood now drying in the outer room, and I split a large log too using a hatchet and heavy hammer found tucked away in the eaves. The log is wet from its life as flotsam, and the inside is eaten with shipworm, now long gone. It has an unusual smell too, like white spirit. But it's drying now, and someone else will enjoy the warmth from it. I have gathered peat, too, from the peat bank cut some years ago, broken it up so it'll dry more quickly, and stacked it neatly for others to enjoy.

Outside this tiny bothy the wind and rain remain incessant, the gusts sound in the rafters and the rain spatters on the fibreglass roof. I'm pleased not to be in the tent, to be honest, in this weather, with several more tent poles broken already; this wind is definitely a tent-breaker.

I've been alone here all day today, and alone yesterday too, and there is no mobile reception here to tempt me. There is solitude here, real solitude. I've been quiet all day. Sometimes I find I'm asking myself a question out loud or reassuring myself that this plan will work or that one will not. I suspect this is quite normal, and it is not like I am chatting away to myself all day. The quiet, the total lack of human interaction, is restful. There is only this place and the goings-on around it to fill my mind. In this weather, though, either looking out for or getting caught by another rain squall, I do get occasional twinges of loneliness. But then the rain eases and there is stuff to do to the boat or the bothy, and looking up at the mountains of Hecla and Beinn Mhòr looming through the gloom at my back I am reminded of the awesome setting. The weather scudding over the slopes, clouds lifting and lowering as they pass through, the bay in front me covered in gusts, flurries of wind pushing water in all directions, and further out amongst the patterns of swell and wind-blown sea, the light playing fantastic. I have not seen the sun all day but out there there are windows in the cloud where sun spotlights white onto the sea, a moving glimmering shine, lovely and wild and momentary. I stare intently at this natural picture show and try hard to seal it into my memory.

Sometimes the colours I see are hard to describe. It seems easy enough when the sun's out. Then, in shallow water where there is sand the sea is turquoise or sometimes a spring green. The contrast then with the many browns and reds of seaweed at low tide, the yellows, whites, greys and blacks of lichen on the rocks, the pink thrift, primary yellow marsh iris, feathery rushes, bracken and heather, a veritable palimpsest of floral progression over time and space, and so pleasing to the eye. These are the Highlands of the tourist postcard, albeit with an extra dash of the untamed in this remote place.

But it is not always like that of course. Much of the time the day is overcast or wet or windy or calm. There were moments during yesterday's paddle when the sea out to the east, with southern Skye and Rum only dimly and distantly visible, was mirror-calm. What colour is that mirror? Does it have a hue of its own? Silver, grey? I don't know. Of course, it mirrors the colours it reflects, and invariably for a sea kayaker that's the sky. But it's never a perfect reflection. On rare occasion I have been out when sea and sky appear to merge, the horizon almost invisible, and were one to look at the view upside down nothing would change. But yesterday the sea was different from the sky – a mirror yes, but still with its own character. Maybe it is the slow-moving liquid, the underlying swell, the viscosity, the depth or my sea-level perspective that gives the sea its unique oily cover. One moment I see it as black, then dark olive, now silver, a glimmer of faint light from deep within, reflected perhaps by some change in density or temperature deeper down. I can't fathom it really, but I can tell you for sure, the sea is not just blue.

It is half ten at night, the light of the day has faded, the fire in the hearth in the bothy has gone out, the temperature is slowly dropping and, with wind and rain expected to last all night, I take myself to bed.

- Route: Uisinis all day, weatherbound
- Distance: 0 km
- Weather: 'A coming as of hosts was heard that was indeed the rain.'

It pleased the wells, it filled the pools, it warbled in the road

It pulled the spigot from the hills and swept the floods abroad

It loosened acres, lifted seas, the sites of centres stirred, then,

Like Elijah, it rode away upon a wheel of cloud.'

Emily Dickinson

· Wind: Red. Force 6+
· Sea state: it's wild out there, offshore westerly, apparently okay from high up, but down at water's level you'd struggle to hang on to your paddle
· Events: time in this quiet place, taking time to watch and listen
· Camping: bothy again.

Monday 4 July, Day 32: Weatherbound at Uisinis, 0 km. Wind: Red.

A day at Uisinis, not just at the bothy but a barefoot walk out to Thomas Stevenson's lighthouse – one of the smaller towers, but built atop a prominent cliff. Not far on the map – but I know that over rough and very wet ground it'll feel like a tough old tramp by the time I get back towards the end of the day. The soil is sodden, and much of the rain of the past few days is still sitting on the surface. It's so wet that I decide to wear my wetsuit trousers, knowing that the squelch will send jets of cold water from between my toes up the insides of my legs.

After a morning spent on chores and a porridge breakfast to die for (oats laced with honey, dates and the last of the UHT cream – such decadence even out here), I set off for the distant lighthouse.

Uisinis is the smallest Scottish light, I think, and is perched on an impressive spot. Lighthouses always feel lonesome to me, built as they were to have people living in them. All the keepers have gone now, and the automation process implemented between 1970 and 1998 has left an eerie silence where once there was the quiet chatter of human activity. The only signs of life today are the occasional throaty 'cronk' of a pair of ravens and the wheeling white flash of gannets feeding out at sea.

The day is lovely. Dry after mid-morning, and though it's still windy the dry is such a treat. There is no way I'd have paddled today; the Force 5 to 7 is manifest in white waves, and the gusts, piling down off the hills, harry the water close into shore. One capsize and I'd be blown out to sea in no time – and of course it gets ever more rough the further out you go.

Returning tired after my tramp, I find I have company at the bothy. Jane, a young lady from England, is taking time out from her cycling journey to visit remote Uisinis. The two of us are journeying in similar fashion, albeit at either end of life. At 19, starting out towards tertiary education, she's spending a gap year exploring and testing her fitness. Clearly deeply intelligent and destined for a degree in all things ecology, she demonstrates an active and mature interest in the issues surrounding farming and land use here in the UK and elsewhere that belie her years. Having bedded into these same issues through my own work for some years, I find we have common ground to generate great chat over several cups of tea. Turns out, too, that she's the niece of one of my wife's walking group friends … small world.

Jane leaves to get back to her bike and cycle on to a hostel. I admire her energy, her determination and her passion both for her journey and for the land use issues she's discovering on the way. She won't be back at her bike before mid-evening, and then she still has several miles to bike to the hostel. Hats off.

Tomorrow the wind is due to drop a notch and stay in the west, so am going to aim for Loch Aineort and possibly beyond. Rain is returning tomorrow and looks like it is going to stay for a couple more days yet, so I need to get myself sorted in a suitably sheltered spot by then. After a half day of no rain I find I've stopped looking over my shoulder; that will change again tomorrow.

- · Route: Uisinis, weatherbound
- · Distance: 0 km
- · Weather: wet at first, then clearing to wind and sun for rest of the day
- · Wind: Red. Force 5 to 7
- · Sea state: gusting off the hills, white water a couple of

Farewell to South Uist

Storm perch on South Uist

Isle Eriskay and a village shop extraordinaire

Escaping the wind, Acairseid Mhor on Eriskay

Orasaigh, Barra – Highland perfection

hundred metres out. Any westerly exposure would be very dodgy

- Events: barefoot walk to Uisinis Lighthouse and return by the old track, now overgrown. Very wet underfoot. Feet feel awesome – a bit sore, but strong – and when I've been paddling the burn-like rash on my upper right foot that I've developed from the rubbing of my shoe has been manageable with a plastic bag inside a sock inside the shoe; peregrines shrieking on the small cliff edges out at Uisinis; ravens; gannets diving out over the white water at sea
- Camping: bothy again.

Tuesday 5 July, Day 33: Uisinis to Hairteabhagh, south-east of Lochboisdale, 19 km. Wind: Green.

Uncomfortable night: foot aching and sore to touch, mind racing, sleeping mat kept deflating, rodents in the roof, and rain almost constant. Had a great fire in the wood-burning stove, though, before bed. Used the drier bits of peat, and after a sooty start eventually got a decent draw. The heat finally dried the kit that had hung wet in the humid air all day.

The forecast is for a window of light winds until 2pm, then rising from the west to Force 4 or 5 and lots of rain. I've formed a vague plan to aim for Loch Aineort today as it's not very far, and I don't fancy a 4am start after such a restless night.

Today's paddle is a delight. The owl, quite likely the same as that seen two days ago, quarters the low ground just above the beach as I set off and everything the Outer Hebrides has to offer is on display. Calmer seas, enough breeze to give a little chop at the open crossings, sun coming and going, bright green hills, water cascading over cliffs, a cave with a waterfall portcullis, Beinn Mhòr and Hecla dominating the skyline, black granite cliffs and fresh-looking rockfalls. And then the wildlife: black guillemots fussing about in a flashing of white, black and red and a whirring of tiny wings, shags galore, peregrine, a lone sea eagle far out, otters, playful seals and ever-squabbling terns. All this set between a blue sea, green islands and blue/white skies and, high up, wispy cirrus forecasting wind to come.

I paddle, stop once to eat and change the map, and soon find Beinn Mhòr falling astern and new vistas of more open country ahead. Loch Aineort comes and goes, and as I'm debating paddling into Lochboisdale to get provisions (an additional 6 kilometres in and out) I happen upon a lobster fisherman in a small boat, hauling pots. A Boisdale man, white-haired and with fisherman's hands, he waves, stops his winch and is up for a chat. Leaning over the gunnel, he bemoans the weather the likes of which he has not seen in many years. The impossibility of fishing for weeks this year so far, pinned by winds and seas, is costing him his livelihood and he's happy to be out today. Two white-haired men in their late fifties, each of us engaged with the sea in our own way, chatting about fish and weather and farming life. He confirms my now three-day-old forecast, and says I've made a good choice in aiming for the sheltered bay south of Marulaig. As we part he offers me a fish – he has plenty of pollock for baiting, he says – and here is the kindness of strangers once again. I paddle away with a considerable-size coley flopping about at my feet. We part with a smile, a wave and a 'good luck', he to his fishing and I to my journeying.

So, Lochboisdale abandoned, I now have fish for two suppers at least. I cross the bay to find the inlet at Hairteabhagh, a great spot, albeit not entirely sheltered from the strong westerlies to come.

I've ended up paddling several kilometres further than planned, and I'm well poised for Eriskay, and Barra beyond, when the winds allow. I pitch the tent on the only flat spot large enough to take it, a tiny tidal island at the head of the loch. This will be home for tonight and likely tomorrow too, so I take time setting out extra guylines. I have no spare poles left.

There are ruined buildings here, houses just above the tide with rounded walls and the suggestion of once housing cattle and people together; a community even here, right out at the end of the world on the southern edge of South Uist. Judging by the state of the walls and what I know from similar settlements across the Highlands, it seems unlikely there has been anyone living here for perhaps 100 years. The tiny island on which I am camped has what could be ancient fish-trap walls extending into the bay, which itself clearly dries a good way out on the low tide. I am never quite sure how these traps might have worked, as they seem such a random method of catching fish. Perhaps when shoals were found on the surface offshore, boats could somehow corral them into the main bay; naturally shaped like the cod-end of a trawl net, it's a perfect shape for a trap, for sure.

I feel dog-tired at landing today; wondering if my body is starting to break down a bit. I've not seen a mirror for a while, but I think I'm quite skinny now; my shoulders feel thinner, and I can see I have no belly at all. Maybe this is how it works; the body loses what it doesn't need, and the strength comes not from large visible 'gym bunny' muscles but from a more efficient use of the deeper core. I am kayaking well, I think, but have noticed just how tired I'm getting when hauling kit up and down rocky shores. Rest is good.

Chasing the forecast, I get the tent up, fillet the fish and gather water from a nearby stream; as I zip up the expected weather front arrives. Blue turns to grey and dry to wet, and the rain sets in once more. I'm safe, though, and dry, and my hope is only that the tent poles will stand up to the battering to come.

- · Route: Uisinis to Hairteabhagh (south-east from Loch-boisdale)
- · Distance: 19 km
- · Weather: sunny, calm; breezy at loch crossings
- · Wind: Green. Forecast 3 to 5, but reckon it was 2 to 3 with some perfect protection on the eastern shore
- · Sea state: calm, breezy on the crossings
- · Events: views of mountain and sea, owl quartering the ground around the bothy, as I paddle away from Uisinis, paddling in the dry, fisherman's gift, unusual campsite
- · Camping: am close to the high tide mark but am fairly sure I have arrived close to high tide, so should be fine, surely!

Wednesday 6 July, Day 34: Weatherbound at Hairteabhagh, 0 km. Wind: Red.

Funny old day. Supposed to be an enforced day of rest. Weather turned fierce as forecast, and a Force 5 to 7 has battered the tent all night and all day today. Incessant rain showers blast through the gap in the hills to the west at short intervals and I have lain awake for I don't know how long, sleeping only fitfully. The steep slopes behind the tent, shrouded as they are in a racing mist, must be providing some protection from the west wind, but it sure doesn't feel like it.

What gets inside your head is the noise. Not the noise of the wind itself, but the rattle and aggressive shake and slap of the tent as it bends to the weight of it. I lay there fretting that more poles were about to break. And what would I do then? I'm miles from any help, and in the almost dark of a misty stormy night, who in this remotest corner of the Hebrides in this remotest corner of Europe was going to answer a midnight knock on the door in any case?

The endless fretting was doing my head in, so it was clear I needed to be self-reliant here. Climbing out of the cocoon of warmth that's my sleeping bag, I stripped off so as not to get yet more clothing wet, and clambered out starkers into the darkness. Headtorch on, I pulled out my rescue line and trolley straps, and manufactured additional guylines for the windward side of the tent. The rain spattered on my nakedness and I cursed myself for not doing all this before retiring for the night. Semi-satisfied, and soaked literally to the skin, I crawled back into the tent, palmed off the worst of the wet from my body and slid back into the now damp sleeping bag to dry out.

The gusts kept coming, fast and powerful. I lay awake, the slapping and flapping constant, and I was on edge. I reminded myself I'm here voluntarily, so need to deal with this rougher part of the journey too. Just over four weeks in, and I think I've had more than my fair share of wind and wet for June. Just like the Boisdale fisherman yesterday, I find this weather bewildering.

With the dawn comes the light – but no let-up in the wind. It does dry up a bit, though, and there is even the occasional temporary glimpse of blue above. It's funny how even such a brief moment like that can be so uplifting. I think – I'm not sure, but I *think* – some of the birds feel it too. They're quiet when the rain is falling or when the wind's thumping across the land, but then with every lull there's a burst of song. Here it's usually the wren or the redshank that I hear, but the oystercatchers will eventually pipe up too – any excuse for a squabble. As the next gusts batter in, the singing dies.

So the day begins wet, continues as such and ends the same. The windward side of the tent is my bulwark against the gale. The extra guylines make a spaghetti of double knots and fishermen's bends and bowlines. The wind has shifted slightly north so my tent orientation is no longer optimal. What to do? Turning the tent round is not an option. This is rough ground, not the manicured setting of a managed campsite. You pitch where you can, you gauge protection from weather where it's available, you put ashore where it's possible, and when weather and sea state require it you live with the conditions as they are and not as you'd like them to be. Compromise, flexibility, mental resilience, physical endurance, being organised, thinking ahead – these are all attributes you need to have in order to undertake such a journey. Chuck in some inventiveness, determination, a well-understood purpose and some positive glass-half-full-type thinking, and you have yourself an attitudinal toolkit for success. I aspire to these, and this trip is testing them for sure.

It's rare to get a chance to journey solo like this, perhaps especially at my advancing age. Family, work, money, duties, community etc can all hold one to a path laid out by others. This is quite normal of course – but I've never been very good at normal. At a relatively young age I met some wonderful people who introduced me to an alternative way of thinking about how to live a life. That, and my reading too: John, Alec, Rog, Ranulph, you and few select others have a lot to answer for. You showed me a different way: that it's possible to veer off the beaten track; that risk-taking is part of the joy; that you mitigate risk with experience; that having a go is better than sitting in too great a comfort at home; and that you need to travel first in your imagination before you can travel for real. What wonders await when you turn imagination into plans and then realities! I've been blessed with some wonderful off-road realities over the years, most of which started with just a sketchy idea of what might be possible. The sketch gathers form as you gather information, colour is added as plans are made, and before you know it there's your painted offering, ready to go. Sometimes you might set off before the framing is done, but that is what having a go is all about. Heading into new spaces comes with the territory, and it's the unknown, after all, that adds frisson.

I have shared some of the most challenging and forming of these experiences with my wife Leah. Our time at Kerracher in Assynt, as just one example, is part of who we are, but we would never have got there without stepping off the beaten track, literally and metaphorically. Kerracher, and our time exposed to the joys and challenges of living and working in a small Highland community, colour not only our sense of belonging, but also our sense of place, where we call home, our politics, our humour, our love of the natural world, our desire to live close to nature and, perhaps most strongly, our sense of community. Those early years, topped up since with our work, the influence of friends, expeditions abroad and some fairly momentous Scottish political changes, have made us who we are.

So here I am now, introspecting a wee bit, pinned by wind in a rattling tent on the western edge of Europe. The open Atlantic is only metres away, and the horizons are as enticing as ever. My mind is full and I have everything I wanted from this trip so far. Moving lightly and with respect amongst these wild places and observing the animals and plants who I share this space with. And I'm testing my own abilities too – my own resolve when times get tough. Thirty-four days in and this is my new normal.

It *would* be nice to have a patch of settled weather, though.

- · Route: weatherbound at Hairteabhagh
- · Distance paddled: 0 km
- · Weather: wet, wind, wind and more wind
- · Wind: Red. Force 5 to 7 from the west, veering about
- · Sea: unpaddleable
- · Events: the noise, incessant
- · Camping: beautiful spot in calm weather, but exposed to westerlies crashing down over the hill.

Thursday 7 July, Day 35: Hairteabhagh to Acairseid Mhòr on Eriskay, 17 km. Wind: Green/Amber.

Last night the wind eased off around 8pm and I started to relax.

The forecast today is for westerlies 4 to 5, which doesn't sound great, but waking to the relative quiet after yesterday's winds is like waking to a calm. I'm keen to get off and on my way, and am on the water at low tide by 9am. The cloud is still low and there's a mist on the water, but it's not pouring with rain – wonderful!

I'm heading south on a calm enough sea in the lee of the shore towards the turn at the south end of South Uist and on to the Isle of Eriskay in the Sound of Barra beyond. Ocean-battered rocky outcrops abound, huge lumps of black granite, deep-cut and steep-sided. At the turn west, as expected, the south-westerly comes on the nose, the sea grows messy and there are a few kilometres of wind-blown exposed paddling up a rocky shore to deal with. Knuckle in, good technique, breathe and eventually you get there. 'There' is the welcome leeward north side of Calbhaigh Island, off Eriskay. The sea here turns from the dark grey of the deeps at South Uist to a gorgeous turquoise hue as the bottom shallows up and turns to sand. Countless billions of shelled molluscs have created such loveliness, and on the beaches themselves, especially the more protected ones, you can find enough intact shells to wonder at the extraordinary variety of species.

I pull in on one such small white beach, drag the boat above the tide, change and walk into the village. The clag is still low and South Uist is now almost invisible in the gloom. But it is still *dry*! What a difference the dry is making to my well-being, and even though it's not actually dry, because the mist makes a moist air, it is at least not the driving rain of previous days.

For a small island, there is a lot going on in Eriskay. Community-owned and once fully crofted, the island's mix of fertilised ground and rough hill grazing sets like a green gem against the shallow turquoise seas of the Sound of Barra to the west. Eriskay ponies – shaggy, white and windswept – roam the hill, and I see a few cattle and sheep. The latter are not as visible as once they were. Just as for the crofters in North Uist, it's difficult for those living here to make a living from livestock alone these days, and crofters everywhere have had to pivot, to innovate, to survive. Tourism has been part of the solution, but with fish farms and fishing also providing employment, and with slowly improving access to digital communications, the way people will make a living here in the future will almost certainly not be the same as it is today.

The Eriskay shop is amazing. That sounds banal, doesn't it? – but this is a shop with a difference. The difference is that alongside the island as a whole it too is community-owned. The ladies running the place could not have been more welcoming or helpful, and the range of stock, when you stop to consider where the shop is situated, is quite extraordinary. I managed to get everything I needed and more, including even – wait for it – dried figs (for a delicious addition to porridge) and a cheap waterproof jacket. The ladies I met are part of a team of thinkers and doers, committed to taking responsibility for making their community into a great place to live. Indeed, most of the Outer Hebrides are now owned by the communities that live there – a triumph of the democratisation of land.

It takes work, commitment and some risk-taking to make community ownership work, and I'm full of admiration for the volunteers who step forward to make things happen. With several years' experience of chairing community trusts and charities as well as engaging with the business equivalent of community ownership known as 'employee ownership' (EO), I have some first-hand appreciation of what it takes to step into these spaces.

Ownership of land does not come easy. It takes dedication, expertise and a healthy dose of resilience not only to purchase the land but then to manage it going forward. Our Scottish government is supportive of community ownership, and there is legislation and money available from the Scottish Land Fund to help oil the wheels. Many words have been penned describing the challenges and triumphs of community ownership, and you can't go wrong in picking up Jim Hunter's *From the Low Tide of the Sea to the Highest Mountain Tops*, published in 2012. Hunter provides not only a historical perspective but also warts and all details of the many community buy-outs that have occurred since 1900. It is an inspiring read.

Community ownership is a success story, but not one without struggle. And that's entirely to be expected. Managing land and the often diverse interests of the people living on it is never going to be simple. The very nature of management by committee means that all sorts of behaviours, attitudes and opinions come to bear. People disagree; some don't know how to behave in a group situation; some misunderstand what it takes to run a business or a trust; and there will always be varied agendas. It works best when, alongside the passion for serving your community, there is relevant expertise within the management team. On the trust I was chair of from 2014, we were responsible not only for capturing and managing some large sums of money from wind farms being constructed in the area, but also for engaging with the community to decide how best to spend it. As a result, we needed expertise in financial management, investment, business planning, conveyancing law, administration, social media, IT security, communications and more.

Community ownership has much in common, therefore, with running a business, and I wonder if there are opportunities for mutual learning between community and employee ownership (EO).

EO is a type of business ownership increasingly popular here in the Highlands, where workers move from being employees to becoming owner/partners and decision makers. Profits are shared within the business itself, not distributed to distant shareholders or to one or two owners. At risk of over-simplifying, in a community land ownership setting EO is essentially the equivalent of the workers on the estate becoming its owners.

My own EO experience was with a salmon and trout processing business based 20 miles north of Inverness, and employing 190 people to fulfil the orders from major UK retail. When we became partners rather than employees we found that the sense of personal stake in the company became palpable and very real. We were a community of our own, bonded not only by making the business sing but also by making it a great place to work. And the ownership element is real. Not only does anyone in the business have a chance to contribute at Board level with full voting rights, but there is also opportunity to broaden your role in the business by getting stuck into any number of group initiatives as suits your interest and capability. Run well, an EO can reverse the traditional, sometimes limiting, top-down management structures and release a new level of energy as more brains are brought to bear on the business needs.

I can only see opportunities in bringing community and employee-owned organisations together for some mutual learning. The underlying themes are already there: democratisation of legal ownership and opportunity, increasing sense of ownership, taking responsibility, releasing entrepreneurialism, creating freedom of choice and more. But there are also particular skill sets relevant to running a business that should be useful to any community that might not have that experience on its team: creating direction, developing people, building teams, managing legalities, project management, providing service, making money, managing finances, using research, and more – all these skills are needed to help make an organisation thrive. So while EO and community ownership may not be a panacea in every case, for those of us lucky enough to have experienced both first hand, they are a wonderful way to work

/ / / / /

I stagger out of the Eriskay shop, rucksack weighed down with food for a couple of weeks ahead and fall into the community hall next door to have a look at the heritage exhibition, which is raising money for a new heritage centre. Local effort, local knowledge, strong sense of place – it's all here. There is mobile signal, so I seek out the forecast. It's not looking great, *again*. The trial of wind and wet continues, it seems, and the wind is due to stay in the south and west and lift to Force 4 to 6. Decisions, decisions; it's all very tricky. Getting out to the islands south of Barra, to Mingulay and Barra Head, is something I'd love to do, but it needs relatively settled weather, especially for a solo paddler, and that doesn't look like it's going to be forthcoming. We will see how it pans out, and I'll continue to poke my nose out and aim to get across the Sound of Barra on Saturday. Tomorrow, Friday, is due a Force 5 to 6.

This is what this journey has been like: taking opportunities when they come, creeping up on open crossings and exposed headlands, and then going for it when conditions allow. I'm thinking ahead three to five days at a time and, to date anyway, there has been no real 'dawdle' time available.

The end of today finds me 6 kilometres south of where I landed on Eriskay this morning. I am camped inside a ruined croft-house on the north side of Acairseid Bay on the south-east tip of the island. The mist is still down and the wind is gusting about, but as it builds during the night I'll be safe behind the solid walls.

- · Route: Hairteabhagh to Acairseid Mhòr on Eriskay
- · Distance: 17 km
- · Weather: low cloud, mist, cold breeze
- · Wind: Green/Amber. Force 2 to 4 and building
- · Sea state: lumpy crossing to Eriskay, then calm under the lee of the island
- · Events: noticing the changing landscape from South Uist to Eriskay; a softening
- · Camping: Acairseid Mhòr, inside the walls of a ruined croft.

Friday 8 July, Day 36: Weathered off in Eriskay, 0 km. Wind: Amber.

Safe within my wind-shattered shell of a croft in the midnight midsummer dim, I slip naked outside to stand awhile, listening to the night. Without clothes, even just without shoes, the elemental nature of living outside is more noticeable. I've been walking and running barefoot now for some years, and that and the wild swimming – or, as we call it here in the Highlands, swimming – has become a regular and life-enhancing activity for me. Something wonderful happens when you immerse yourself in cold water or when you tread the mountain without shoes. Your skin tingles, your heart beats hard then settles, the soles of your feet become supple like leather, and the body feels light, free. There is the science – proprioception and endorphins – to rush in here if you want it, but I just love the simple feeling of it, the opening up, the immersion.

The cloud is still low, and the wind, gusting fresh, blows around my skin and stirs the sea loch beneath. Out from the small seal colony on the other side of the loch, mournful cries carried on the breeze remind me that this is still a wild place with creatures living beyond our view: the guillemots perched on cliffs, their young fledging and ready for the leap into the void; the storm petrels, tiny black birds of the night, huddling under rocks waiting for their partners to return from the sea with food; mother otters teaching their pups to fish; dolphins moving through the dark offshore, their puffs and blows giving them away; and of course the seals, calling to each other through the dark. A whole world of lives happening beyond us, despite us, out of sight and thus too easily – fortunately in the short term for those creatures, not so fortunately in the long – out of mind.

As dawn comes the seals continue their plaintive cries. Their perches, high off the water on the seaweed-covered rocks, are fast being encroached on by the rising tide and soon the animals will be back in the water. They survive where I'd be dead in minutes. The Arctic terns have risen squabbling from their offshore skerry, and at their passing a Jenny wren has just sung out from the wall by my tent. Such beauty from such a small being.

Resigned to not paddling today, I take a walk back into the village, the *baile* as it's known in Gaelic. Strong westerly winds are blowing across the Sound of Barra, and the view west, out over the turquoise sound from the high point on the single track, is to die for. White beaches, distant islands, blue seas, well-kempt houses dotted over the green land, crofts overrun with vegetation, the machair in full bloom with a perfect palette of yellows, pinks and greens. Eriskay ponies are grazing in a fenced enclosure, and high up on the road a statue of the Virgin Mary surrounded by benches awaits Sunday's outdoor congregation.

In the pub, the Am Politician, all is a-bustle. The place is famous for its name, of course, and the story of the shipwreck and local snaffling of bottles of *Whisky Galore* by Compton Mackenzie is part of the folklore of these islands. But the pub is known, too, for its welcome, its conservatory views out over the Sound of Barra and, today at least, for its delicious cranachan.[8]

Tomorrow, Barra, weather permitting.

- Route: weatherbound on Eriskay
- Distance: 0 km
- Weather: wet then, dry from mid-afternoon; strong winds all day

---

8 A glorious assemblage of oats, raspberries, honey, thick cream and whisky.

- · Wind: Amber. Force 5 to 6 from the west
- · Sea state: big lumpy water from the west – no time to be out alone on the south side of the Sound of Barra
- · Events: Am Politician, total treat
- · Camping: Acairseid Mhòr, inside a ruined croft-house.

Saturday 9 July, Day 37: Loch Acairseid Mhòr to Castlebay, Barra, 21 km. Wind: Amber/Green.

Restless night, nervous about the crossing today. Wake very early, footle about, and get on the water as planned by 7am. Aiming to cross the Sound of Barra around the bottom of the tide so as to minimise the potential mess that is the flow of water versus the wind.

My timing is good, the weather window holds and I cross the gap between Eriskay and the outlying islands of Gighay and Hellisay, offshore from Barra, with a bit of lump and some tidal shoogle in parts. The gap feels wonderfully exposed, with open sea vistas to the east and west dotted with islands near and far. At Gighay I'm greeted by a sea eagle taking flight just a few metres above my head. Flapping hard, it hauls itself heavily up the slope and perches on a rock at the top to watch my passing. As I come round the southern end of the island, it's still there, flapping from rock to rock and apparently following me round. These twin islands are part of an ancient volcano, and it's tempting to head through the gap and into the drowned cauldron – but with the wind due to freshen early afternoon it's not the day to dawdle.

Paddling on, leaving the caldera astern, making smaller crossings between other islands, wind on the starboard bow, passing salmon farms and exquisite groups of skerries all golden browns above and turquoise beneath, and coming to rest eventually on Orasaigh, a tiny almost-tidal island on the east side of Barra, off Earsary.

And the sun comes out! I decide to ignore the freshening breeze to come, and instead give myself a treat by stripping off and swimming in the cold crisp crystal-clear waters of the bay. My kit is laid out to dry on the rocks and I have a full body wash, a tonic to do Mr Thoreau proud.

Within an hour the mist has returned, and one exposed windward coastline later a tired paddler emerges beneath Kisimul Castle in Castlebay, the capital village of Barra.

Barra, a gorgeous big rocky lump right at the southern tip of the Hebrides and fringed with white beaches to die for, is home to over 1,000 island folk working as tourism providers, fishers, fish farmers, processors, healthcare workers, engineers, builders and more. Tourism is the mainstay, though, and in the summer the ferries from Oban and Mallaig bring a steady stream of visitors here, into Castlebay in the south of the island. I paddle by the restored medieval castle of Kisimul guarding the bay just off from the ferry port and wonder, with the thick clag that often shrouds this part of the island just as it is today, how it is that Calmac have not yet crashed into it.

I've been travelling to Barra for years, sometimes with work on the salmon farms but also with family and with friends, and it feels great to have got here, right to the almost tip of the Hebrides after such a long journey. I say 'almost' because south of here, beyond the causewayed Isle of Vatersay, is a small string of uninhabited islands culminating in the towering Mingulay and Berneray, 30 kilometres away. From there, if you sail just west of south, it's open ocean all the way to South America, and the 4,500 miles of fetch can make for some big scary seas.

I have kayaked through those southern isles before, but with a forecast of Force 5 to 7 for a few days ahead it's not looking possible to do so this time. So I book myself onto Monday's ferry to Oban, paddle out to a quiet shore on the edge of the village, heft my gear up on a rough piece of grass and camp up.

Enter Donald, a local lobster and prawn fisherman, now ashore for the day and tying his boat on to a running mooring close by. We chat for ages and he urges me to attend the Barra boat festival tomorrow, Sunday. This wonderful local tradition, when the island priest holds a Mass and then splashes holy water on to the fishing boats to keep their owners safe, has not been held for two years due to Covid. This year's festival is thus eagerly anticipated by all.

- Route: Acairseid Mhòr to Castlebay on Barra
- Distance: 21 km

Kisimul castle off Castlebay on Barra

Boat festival bunting on Barra

Barra boat festival

Castlebay chaos, Barra boat festival

- Weather: an overcast and misty start give way to sunshine then return to mist
- Wind: Amber/Green. Breezy start then drops to around Force 3
- Sea state: bit of lump in the sound then calm in the lee of Barra
- Events: swim at Orasaigh
- Camping: west side of the village of Castlebay, overlooking the castel.

Sunday 10 July, Day 38: Attending Barra Boat Festival, 0 km. Wind: Green/Amber.

I don't paddle anywhere today but instead settle in the village and focus my attention on the Barra Boat Festival. I'm in for a treat. At lunchtime, an armada of boats bedecked with bunting begins to motor in and tie up *en masse* around the piers. The harbour is similarly decorated, and an open-sided lorry container, reformed into a Catholic altar, sits at the centre of festivities in the ferry car park.

People start to gather. Locals and tourists, fisherfolk with their fishing boats from all around Barra and southern South Uist, the Barra lifeboat and crew, dignitaries in suits, a pipe band resplendent in MacNeil blue/green kilts, mothers, toddlers – and even teenagers, the latter conspicuously enthusiastic over this very local celebration of what the sea, and the fishing in particular, mean to this community. With everyone milling about the harbour area, the festival launches with the pipe band leading the gathered throng through to the altar platform; three Catholic priests, their helpers and an altar boy or two, white cassocks flapping in the breeze, follow close behind. A rather dour service follows but is enlivened, rightly or wrongly – probably wrongly – by the not-so-distant sounds of laughter and merry-making from the fisherfolk who have chosen to remain on their boats. The contrast of the priests' droning voices and the occasional distant peals of laughter from the pier is kind of wonderful.

With jugs of water suitably blessed, the priests move off to the harbour to sprinkle – actually, more 'spray' – each boat with the now holy water as a blessing to keep their fisherfolk safe out there on the open seas. In receiving the blessing some bow their heads in reverence, others look a little bemused, and one wonderful lady, clearly full of a dram or two, opens her arms wide to a somewhat embarrassed priest and shouts, 'Come on, then, Father, give it to me, why don't you?'

This festival feels local. Yes, there are some tourists and I'm one of them, but we look to be in a minority. The overwhelming feel of the day is one of family orientation: kids running about, folk in working clothes, white heads in the pews, other less mobile elderly sitting in their own bus with the windows open to hear the ceremony, mothers with babies in arms or prams, fishermen in yellow wellies and filthy t-shirts seemingly oblivious to the chill in the misty air, chat, laughter and, yes, later on, the increasing anarchic din of folk overindulging in the demon drink. What is a festival, after all, without a party or two at the end of it?

But before the evening parties there's the traditional armada in the bay. You need to be there to really get a feel for it and my words will hardly do it justice: chaos, fun, whoops of delight, fishing boats packed with young people, engines gunned, orange flares, horns, white wakes, zero safety, free-spirited, anarchic, sunlight burning off the mist. With everyone going a bit crazy out there in the bay between Barra and Vatersay and with big powerful boats criss-crossing at speed often just metres from each other, it's a miracle no one gets killed. But no one *is* killed, and I love the informal, devil-may-care and inclusive atmosphere of it. A real letting-off of steam, and a reminder to those watching that there are still communities here in our islands where the sea looms large.

Oh, and did I mention the free cooked herring? Delicious!

Tomorrow I'll head away from these islands to continue my journey north. Spending more time here would be my preference, but the weather, as ever, seems intent on moving me on. Kayaking through the Hebrides has thrown up such a variety of land and seascape, and the journeying has been far from simple. Wind and lumpy seas have been ever present, and I've spent a lot of time living with uncertainty, keeping a weather eye for the next rain shower or the next squall.

But that same uncertainty has thrown up some wonders too. Alongside the ecology and sensual input there has been much to think about around community, land ownership and how these islands might hold onto their people and their sense of place, and avoid becoming just another tourist hub. Here, in the southernmost tip of the westernmost edge of Scotland, the Barra Boat Festival is an example of a lasting local tradition which contributes to that sense of place. Kept alive by the young and not so young alike, the turnout, the excitement and the passion with which so many have engaged with the day's events suggest that this celebration of seafaring remains relevant and culturally significant. I'm biased, of course, wedded as I am to the sea, and I want it to be so – but I'm aware too that island living, beyond the tourists' rose-tinted gaze, can be, to say the least, challenging.

Islands by definition are bounded by the sea, but being bounded by it doesn't need to mean being *bound* by it. Not today, in the 21st century. Ferry services, flights, IT networks, social media, our natural capital, all these throw up opportunities for islanders to engage with the world at large. Horizons, wide though they may be here on the Atlantic edge, can now be extended to markets unthinkable to those living here even 20 years ago. And as skilled working and home working continue to grow in popularity, so will the ebb and flow of islanders change the faces found in the pubs and shops. Change is inevitable and normal, and it is perhaps speeding up as new technologies make the previously impossible possible. Within it all, I hope the sense of place I have seen today, this deep connection of people with the land and the ocean that surrounds them, will endure.

The ferry is to depart at 7am tomorrow, so in anticipation of the forecast big blow from the south in the morning I paddle my kayak over to the jetty this evening and leave it there overnight.

- Route: stayed for Barra Boat Festival
- Distance: 0 km paddled
- Weather: misty, breezy in parts, some sunshine
- Wind: Green/Amber. Breezy in parts
- Events: Barra Boat Festival, armada, revelling late into the night
- Camping: west side of the village.

Monday 11 July, Day 39: Ferry from Castlebay to Oban, meet Leah, dry and repair kit. Wind: Green/Amber.

At the jetty this morning a group of five kayakers appear with their boats on trolleys, ready to board the ferry. They had travelled out on Thursday and attempted to get out to Mingulay, but wind and waves had meant they had only got halfway to Pabbay before turning back. Disappointed but in good form they invite me to join them for breakfast on the ferry. Great craic[9] and serious discussion too, and with three ex-teachers, a special needs nurse and a youth charity person in the group educational issues loom large alongside tales of kayaking derring-do.

Leah. looking as lovely as ever, is at the quayside when I arrive in Oban. We drive out to Dunstaffnage, park up by the castle overlooking a bay filled with yachts, and festoon the trees with my kit to dry it off. SAMS, the Scottish Association for Marine Science, are great too, in giving me the use of a hose to rinse off the salt. I sleep well in clean sheets in a B&B for the first time in three weeks; such luxury.

- · Route: ferry to Oban and sorting kit
- · Distance: 0 km paddled (ferry to Oban)
- · Weather: dry, breezy, sunshine later
- · Wind: Green/Amber. Force 2 to 4
- · Events: meeting fellow kayakers on the ferry, drying out at Dunstaffnage
- · Camping: B&B

---

9 Great chat, good wit and fun company.

6 | Midge lover

Days 40 to 45 – Oban to Glenfinnan to Uags, Applecross Peninsula

Tuesday 12 July, Day 40: Glenfinnan to south of Gaskan (north shore of Loch Shiel), 17 km. Wind: Amber.

After 40 days' challenging journeying on the sea more of my kit is starting to wear out, and repairing things is becoming a daily task. I spend the morning making essential purchases, getting my washing done and sorting food and kit, ready for the final three weeks of this venture.

The route ahead will be guided by weather, as ever. The winds have been staying in the west for days now and are forecast to remain strong, at Force 4 to 6, for most of the next week. During this trip there is no sign of the high pressure area I might reasonably have expected – and long for. Decision making therefore continues to be complicated and marginal. Most of the coastline north of Oban is fully exposed to the west, so all I can do is make a plan based on what info I have available, and flex and adapt as required in the days to come. For now I choose to head overland to Glenfinnan, as it will allow me to continue journeying through the Highlands, and from Glenfinnan I can go back to the sea via the loch and the river without needing another lift. Hopefully, the weather will change by the time I emerge from the river, and permit me to make progress north. It's all 'hopefullys' and 'maybes', but that's been my normal for some weeks now and I'm kind of getting used to living with the uncertainty of it.

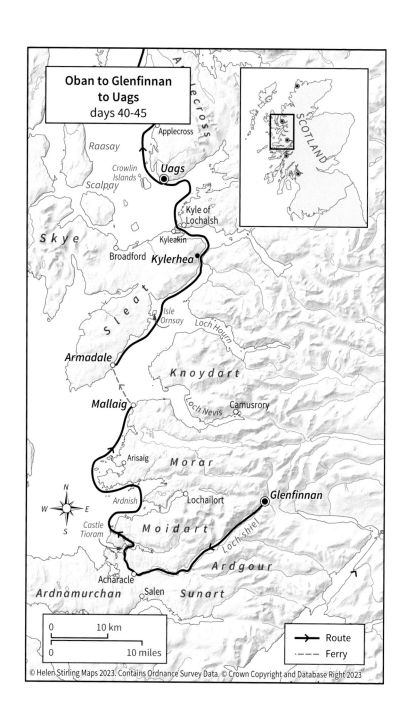

**Oban to Glenfinnan
to Uags**
days 40-45

SCOTLAND

Applecross

Raasay

*Crowlin
Islands*

Uags

Scalpay

Kyle of
Lochalsh

S k y e

Kyleakin

Broadford   **Kylerhea**

*Isle
Ornsay*   *Loch Hourn*

S l e a t

Armadale   **K n o y d a r t**

Loch Nevis   Camusrory

Mallaig

Arisaig   **M o r a r**

Ardnish   Lochailort   **Glenfinnan**

N
W — *E*
S

*Castle
Tioram*   **M o i d a r t**   *Loch shiel*

A r d g o u r

Acharacle

A r d n a m u r c h a n   Salen   **S u n a r t**

0	10 km
0	10 miles

→ Route
--- Ferry

© Helen Stirling Maps 2023. Contains Ordnance Survey Data. © Crown Copyright and Database Right 2023

Glenfinnan and Loch Shiel, famously the site of Bonnie Prince Charlie's rallying of the clans for the 1745 Jacobite rebellion, are gorgeous places to be, in any case. Cutting deeply between steep-sided mountains, the loch is 100 metres deep, on a cliff-like gradient. It's a ravine; take the water away, add a winter's snows, chuck in a condor or two, and the vista would match any Andean spectacular. I'll be paddling against a Force 4 to 5, but in the relatively sheltered inland water, so with a bit of graft I should be able to creep forward. Tomorrow the wind is due to lift to Force 5 to 7, which will make the second half of Shiel challenging indeed, but even if I get pinned, at least I'm still out here, engaging with the wild places and not sitting feeling sorry for myself stuck somewhere on a westerly shore.

Once down the river and through Loch Moidart, I will have options: turn left or right. Left points to Ardnamurchan and a potential crossing to Muck and the other Small Isles, all of which will be very weather-dependent. Right points north, exposed to the westerly wind and still at risk of getting weatherbound, but aiming for Arisaig and north from there. I won't need to make that decision until I'm in Moidart, though, so for now I'm settling to a long grind against the wind blowing up Loch Shiel.

Leah drops me off at Glenfinnan late afternoon. I paddle for three hours in a wetting short chop characteristic of freshwater kayaking, and find a spot on a shingly beach south of a wooded area called Gaskan. The wind begins to ease as I come ashore and the rain holds off while I get sorted – and then comes on again in earnest all night. Once again I lie awake listening to water sounding and wind rushing about the loch. All is a-tumble in the dark, the waves splashing ashore a couple of metres away, the trees thrashing about just above my head and the rush of wind rattling the tent.

- Route: Glenfinnan to Gaskan
- Distance: 17 km
- Weather: dry! – in the daytime, at least
- Wind: Amber. Force 4, gusting 5 on the nose
- Sea state: fresh water, short choppy
- Events: loch feels eerily quiet of wildlife; one heron, a goosander and a lone gull; a bit lonesome

- Camping: shingly beach; no tide, and little risk of the water rising at this time of year in this size of loch – so camp just above the water.

Wednesday 13 July, Day 41: Gaskan to Castle Tioram in Loch Moidart, 17 km. Wind: Amber.

I am halfway down Loch Shiel, bedded deep into this precipitous landscape. Low mountains, but mountains nonetheless, rear up behind my tiny camp spot and in all other directions. Everywhere I look is steep, gnarly, rocky – and green, so green. The slopes low down are painted with bright deciduous and dark coniferous trees, bracken shrouds any gaps, and above the dwindling forest, coarse grasses and mosses carpet the slopes. The loch in between is a long thin stretch of blue or grey, white-flecked as the wind rises, and whenever the wind lulls and the water is still, the sky above it all is mirrored in its surface.

I am sat, leant back against my boat, togged up against the breeze and basking in the occasional burst of sun as the fluff of clouds above is blown east and north. It's gorgeous – and there's no rain!

For me, I think it's the water that creates the beauty in these landscapes. Ever moving, restless, a mirror of wind and sky, the light glittering on the tops of wavelets and smoothing to a silver sheen further out. As the wavelets break, they cast a momentary black shadow out below the light, and in between those shadows, the watery slopes glimmer a fluid aluminium sheen. The detail, when you take time to look, is stunning. When, on occasion, the breeze drops I notice the insects rise. Long-legged mayflies, the occasional bumblebee and of course midges, countless billions of midges.

We moan about them – but where would we be without the midge? Fewer flies, fewer spiders and other insect hunters, I suspect; then fewer birds, fewer frogs – and less tang of midge spray in the mouth and eyes, less to complain about. But also, possibly, more tourists, more holiday homes, more human imposition on the landscape. Perhaps – no, there's no 'perhaps' about it – we need to learn to love the midge. This humble beastie, with mouthparts which if viewed under a microscope would send you running for the hills, is both scourge and saviour. It torments as a cloud, descending on our camp, our barbecue, our quiet contemplation out of the wind, and it can drive us to distraction. Out here of course I could run for the tent and zip up, and sometimes I do, but mostly now I will put on long sleeves, squirt some Smidge, move about, find a breeze and remind myself that the rough and the smooth are partners in creating all this gorgeous surrounding landscape.

The beauty I so enjoy out here lies not in perfection but in variety and duality, I think. If the sun shone all the time, the pleasure of it would wane. But when the sun appears after rain, it is the contrast, the relief if you like, that is so pleasing. Dualities abound everywhere we look, of course, and when you think about them in terms of our wilder places here in Scotland, they both add to the challenge of moving through this place and, then the next moment, provide relief too. Up/down, steep/shallow, stormy/calm, hot/cold, soft/hard, wet/dry, natural/unnatural, unsafe/safe … midgy/not midgy. That calm after the wind, the sun after the rain, the easing of the slope as you near the mountain ridge, the light breeze that sends the midge diving for cover. The joy of all that is in the contrast, in the change.

So I have learned to love the midge. Well, 'love' is far too strong, and maybe 'accept' is more honest. The midge's place in our ecosystem is well enough understood, and central to it. Alongside other wee beasties, it is akin perhaps to the krill in our polar seas. Its baseline can support countless other creatures that in turn are food for others. We might not want the midges when we're camping, but in the wider scheme of things we'd be the poorer without them.

As I sit here writing, the breeze has strengthened. But the sun appears more often now, and huge cumulonimbi are lumbering across the sky. In amongst the shingle around my feet, I am watching wee beasties dotting about; small black spiders no bigger than my thumbnail are skittering between the no man's land of shingle and the small hollows and rises made by the feet of the deer as they walk along the shore. The latter give away their presence by leaving their footprints, and their pellets too. Perhaps this spot is a favourite watering hole, or just an open space as relief from the tight bounds of the dark woodland above. These spiders, though, they move with intent. A dash here, then stock still, a wee turn there and back, another dash. I have not yet seen a pounce and a kill, but it must come. They move in and out of view with every dash, blending as they do with the darker pebbles. Ants too are crawling onto my legs, tiny and black and ever searching. It is a wonder to me how they navigate through this convoluted landscape and still get back to the nest: a spectacular feat of spatial memory and search techniques. A footstep for me is several horizons for these tiny creatures. And they are so robust. I have one now, running over the page I'm writing on. It has reached the edge and the wind has lifted the page and flung it back onto my lap. Straight back on to the page it goes, seemingly unharmed, only to be thrown off again. How do they survive such a violent act? It's the equivalent of me being thrown the length of my house. I would land crumpled and broken, but these tiny beasties bounce and roll, their pin-thin legs unbroken, before they leap upright and are away. The answer lies in their relative lightness, their exoskeleton perhaps, and their innate 'otherness'.

But back to the midge. I have suggested that it's a saviour as well as a scourge. I wonder if it's a saviour because it gives the Highlands a protective sheath against overexploitation by that pet concern of mine, rampant tourism? Now that sounds a little floppy and perhaps even a bit controversial and anti-tourist – which I am not – but bear with me. Sun, sea, beaches, heat, easy living, these are what seem to draw many of us on our hols. Ever more of us seek these out by flying to the costas rather than holidaying at home. Scotland too – and I'm talking about the Highlands here, not our towns and cities – has spectacular world-class scenic splendour which includes the sea, beaches and occasional sun. Right here, on my wee pebble beach halfway down Loch Shiel, I'm surrounded by everything Scotland's tourist brochures promise. And yet there's nobody else here! Why? Well, because as well as being a beautiful place, it's also a difficult place. The wind is fickle and often strong, the temperature varies, rain is seldom far away, and the midges can be absolutely horrendous. So, if it's easy, pain-free vacationing you want, do not come to Loch Shiel – not down the loch, anyway.

I worry about this stuff because, with the benefit of more than 35 years' living and working here, I can see that certain forms of tourism seem to be speeding up change within our Highland landscape and communities, and not always for the better in my opinion. If you need an example of where we're headed, go to a village on the coast of Cornwall in the summer. No, that's too obvious. Go in the *winter*.

Go in the winter, and count the blackened windows at night. Easier, actually, to count the lighted ones. The blackened windows are the holiday homes, often second homes, owned by people who don't live there or indeed anywhere near there in the main. The big money from the cities is buying up – buying away in my opinion – the heart of many of our coastal communities. Cornwall's beaches, relative high levels of sunshine, apparent quaintness, and no midges, have made its coastline a magnet for the wealthy get-away crowd.

And the same thing is happening here in Scotland, in the Highlands too. The landscape and some of the communities through which I am journeying and which I am engaged with in my work, are starting to look a bit like the Cornish model. Young people local to the area, teachers and doctors moving in, fish farm workers and others are resigning themselves to having to rent, some perhaps for all their lives. This is not good enough. I have spoken to very many local people over recent years from all over the Highlands, and the message is always the same: 'It's the holiday and second homes, they've priced us out of the market.' On Mull, in Arnisdale, at Port Gobhlaig, anywhere on Skye, Ullapool – pick a community, any community, and the housing stock is too often heading to holiday/second homes.

And the result? Well, it's a kind of clearance. That's an emotive image I know, but when people can't afford to live somewhere they move away, and in doing so they take with them their youth, their energy, their children from the school, their labour from local business, in many cases their language – and, just as important, their sense of place. Instead of the welcoming homesteads are houses left empty for much of the year and, if rented out as holiday lets by owners who don't live locally, profits that disappear to wherever those owners are living. Yes, such a holiday let *can* bring in some cash to the local community in the form of a bit of cleaning work in the summer, food purchased locally (assuming the visitors haven't bought all their food from the big city supermarket in advance) and occasional guided trips to local beauty spots perhaps. But, compared to the core business profits, these peripheral spends are crumbs from the table, and distant owners sure don't add to the lived-in local community life.

And as the second home plague spreads and there is no longer anywhere for the providers of local services to live, who will provide

Loch Shiel

Sparse camping on Loch Shiel

St. Finan's isle, Loch Shiel

Western Loch Shiel from St.
Finan's isle

13th Century Castle
Tioram, Loch Moidart

the services for the holidaymakers? If the owners are also living in the community I'm less critical. Then at least the profits remain local, along with the personal social engagement which goes to create community itself. But individuals buying up several homes in Portree or Poolewe or Port Gobhlaig and converting them to holiday lets – that is not so good.

We love our Highland places, but we seem to be really adept at destroying the things we love. Oscar Wilde explores the theme in his poem 'The Ballad of Reading Gaol'. His observation 'For each man kills the thing he loves' relates to a man whose jealous love of his wife led him to murder in that, as we infer, he would rather have her dead than potentially available to others. Wilde's story seems extreme, but his message is universal. It is an observation on love, obsession and a need to own, to control the object of one's desires. Chuck in the money to be made by selling houses for second homes and a social media frenzy, 'my top ten secret Highland places', and we have a seductive motive and process for picture postcarding our landscape and selling the images off to the highest bidder. We are in danger of loving our landscapes and communities to death, and I, for one, think we need to act in an entirely different way to avoid killing this thing we love.

Maybe the midge plays a part, maybe not, but thank goodness for it and for our fickle weather. The latter has been especially prevalent this year, 2022, it seems, for unlike the heatwave south of the Highland line, we have had a succession of lows with high winds, plenty rain and nothing settled for weeks. It's been a brave tourist who's visited us this year. The weather and the midge – our saviours!

So rampant tourism needs to be checked and better managed. Not all tourism, but those elements at least that are emptying our communities of their localness; and 'local' doesn't have to mean born here – but it does need to mean *live* here.

The way we create lived-in communities is by promoting and encouraging other types of business, other entrepreneurs to create the jobs that sustain our Highland life. Let's start with one that has been a great success story in recent years, and one I've seen much evidence of during this journey: the farming of Scottish salmon. Controversial to some, misunderstood by many and occasionally intentionally misinformed about by a few, salmon farming is one example of such entrepreneurialism. Using the natural capital of our seas, our lochs and our rivers, salmon farmers operate in an economic environment that pays well and does so year round. The industry is not perfect, and like any relatively new business it has challenges, not least around its environmental impact. But all industry, i.e. every time something is made or a service is offered, impacts either directly or indirectly on the environment. Ultimately it's a trade-off between impact, companies' value systems and market desire for the goods on offer; salmon farming and tourism are no exception. For the former, which I have some first-hand knowledge of, I take heart that the impacts are at least well understood, and now increasingly well regulated both by legislation and by discerning customers' purchasing power. The industry is not there yet, but there is a direction of travel that feels positive. Over and above the year-round employment, salmon farming builds transferable skills, encourages education, grows food, develops young people, and keeps workers and their families in the area. All of that is a far cry from the very seasonal, generally low-skilled and often low-paid jobs that much of tourism provides – with some positive exceptions, though, just as there are to any rule.

So we're not Cornwall yet, but you could argue that Skye is rapidly moving that way. The midge is doing an excellent job of keeping the reality of our beauty, our innate dualities, alive in the minds of many who might otherwise keep buying up our housing stock and blackening our windows. But the midge cannot do it alone. Government intervention is needed as well, to protect our communities. We need to build affordable houses, and cover them with enough regulation that they cannot be instantly sold off for personal profit as holiday homes but will remain as local homes. Community housing trusts have a significant role to play here. And we need robust taxes on second homes. And I am talking *properly* robust – not £10k here or £20k there. If you can afford a £300k second home then £20k is not going to deter you. So in these sensitive areas we need to go much further. The tax should be prohibitive – that's the point, isn't it? And if we decide this is too draconian, then we could perhaps add some planning rules to protect or reserve properties for residential use.

That in turn might include allocating areas specifically for tourist properties or perhaps copying parts of the Norwegian *boplikt* system, in which an owner is required to live year round in any residential property they have purchased.

And, supporting these moves, as already explored back in Eriskay in the Outer Hebrides, we should keep encouraging community ownership, too.

/ / / / /

Sensing a change in the day, I look up from my notebook to see that the wind has dropped a notch. I feel it first, then watch as the white waves give way to blue. Time to go. An hour to pack everything away, and the rest of the afternoon goes something like this:

RAF Hercules flying low over the loch as I set off; primary colours as the sun holds forth; wind on the nose all afternoon; warm enough for paddling in t-shirt; St. Finan's Isle with its ruined medieval chapel and ancient crosses; two fishermen in a boat on the windy loch, lost fish, snapped line; grinding it out against the breeze; the vast expanse of the Claish Moss raised bog to the south; weedy shallows; sense of coming out of the wilderness at Acharacle; moving into the Shiel River, dark, slow flow; fishermen's good-natured chat; tide at the river mouth caught perfectly, high water, paddling straight out to sea, barely a ripple; world opening up from the mouth of the river; lumpy water from the open sea to the west; and finally, a long day done, camping on slopy wet ground beneath the 14th-century Castle Tioram. Sweaty night, nervous about tomorrow's windy exposed section to Smirisary and beyond.

- Route: Gaskan to Castle Tioram
- Distance: 17 km
- Weather: dry, sunny!
- Wind: Amber. Force 4 to 5 on the nose
- Sea state: freshwater short chop, but mostly just wind; back out at sea, the westerly has clearly built up a running sea which lifts my keel as I paddle out from the river to Castle Tioram
- Events: the peace of the loch, writing journal outside,

> sun-kissed paddle cooled with wind, St. Finan's Isle, the river mouth slack with high tide
> · Camping: tidal island at Castle Tioram.

Thursday 14 July, Day 42: Castle Tioram to west end of Ardnish Peninsula, 17 km. Wind: Amber.

A day of contrasts. I want to catch the high tide not just to get off the tidal island that is Castle Tioram but also to get west up the north channel of Eilean Shona, which itself dries out at low tide.

Up at 5.30am and on the water an hour later. What an hour, though! Pouring with rain, and I mean pouring. It doesn't start until I've started pulling the tent pegs, but then it just goes for it. My new £1.50 plastic poncho from Oban does the trick in keeping me reasonably dry (my kayak top is leaking copiously now) but the tent and all my drybags get drenched. It's a test.

As I get onto the water the rain passes, and I lean into the stiff westerly breeze up the north channel. Steep, thickly wooded ground rising on both sides – and then at the mouth of the channel the world opens up. Emerging from the relative shelter, I have a sudden feeling of space, of release from the tree-clad channels and muddy shores of Moidart, and the open sea and sky widens to the west. It's a real seascape. The Isle of Eigg and its pointed volcanic plug of An Sgùrr are blue in the hazy distance, the sun is putting in an appearance every now and again and turning a glowering sea shiny blue, breakers are painting the shorelines of the outlying islets white, and the long peninsula of Ardnamurchan, that 'headland of the great seas' in Gaelic, stretches away to the south and west.

Out of the channel the sea is lumpy, so I sit in the lurching water, wind gusting about me, taking it all in and assessing what next. I have reached that turning point: left towards Ardnamurchan and the Small Isles, or right towards Arisaig and beyond. Now, though, the decision is easy: the westerly is blowing Force 4 to 5 and the sea is as big as I've seen since Toe Head. West and into the open water of the Small Isles is very obviously off the cards, so I set my bow north.

All is sunshine and wind, and the sea is big and messy. Keen to keep moving, I take a deep breath and pull out into the waves. Now with

On the edge in Ardnish

Rough bounds east of Ardnish

West to Muck, Eigg and Rum

them on the beam as I come north of the tiny Eilean Coille the reason why paddlers say 'watch out' when discussing this small stretch of coast becomes all too clear. I quickly find myself in uncomfortable water, bracing far too regularly and nerves starting to jangle. Here, the wind has a long fetch to the west and it's been blowing hard for days. That, combined with shallowing water and a flat rocky coastline, conspire to create a tricky mess of steep waves, some breaking, and lots of jubble from the clapotis. I throttle back, take heed of my gut and run away, slowly turning round to keep my balance and heading back to the lee of the island. It's too lovely a day to smash boat and body on the rocks.

Opposite the island, on the mainland, there is a perfect crescent of sand with machair above it; a really unusual feature in this part of the coastline and well sheltered by the island. It's a godsend on this day in this wind so I land there with some relief. The sun is still shining, so I use the time to stretch the tent out to dry, make tea, reassess, wait and watch the weather for a lull. Even a small drop in the wind would do it. And so it turns out. After a couple of hours, the edge goes off it and the wave height looks like it's dropped a little. I take a walk out over the rough ground to the headland to see what it looks like, return to the boat, and go for it.

So here I am, ashore again, several hours on, feeling a little frazzled from what was a rather uncomfortable paddle in a confused bit of sea. But I'm feeling somewhat self-satisfied, too, having taken it on, especially alone, and am now perched on a tent-sized bluff above a tidal white beach on the western end of the Ardnish peninsula. To the east, round a couple of small headlands, is Peanmeanach Bothy, the so-called kayakers' bothy, but I'm choosing the solitude of my tiny bluff over the chat of the bothy, and in any case I've heard that the owner has closed it to visitors. The forecast is for double raindrops all night and half of tomorrow, so I take some time checking the tent. Swim, wash in the burn, feed, read, write, doze, listening to the crashing surf just below the tent now the tide is high, and remove a score of tiny ticks from my legs. I think I'm camped on top of a nest somewhere.

- · Route: Castle Tioram to western end of Ardnish Peninsula
- · Distance: 17 km

- Weather: wet early then dry until evening
- Wind: Amber. Westerly, Force 4, dropped to a 3 for a couple of hours
- Sea state: unpleasantly rough, short chop and gusty wind making for some uncomfortable paddling
- Events: edging out of Loch Moidart, drying kit on crescent beach, rounding Smirisary, afternoon swim
- Camping: exquisite spot on a bluff with views west, sparkling sun on the water, waves crashing just below the tent on the high tide.

Friday 15 July, Day 43: West end of Ardnish Peninsula to west end of Arisaig Peninsula, 13 km. Wind: Amber.

Poured with rain all night, truly *poured*. Tent held up and I slept okay. I awake to the rain clearing and the sun poking out of a cloudy sky. A dry forecast is to be savoured. However, the wind is still a bit full on, Force 4 to 5 from the west, and out beyond the gap between island and mainland the sea is white with breakers. With tide moving against wind this morning, shooting the gap looks intimidating and, as I watch, a fierce running sea is lumping up at the point.

Just to add to the pressure, I'm also being attacked by slugs. Admittedly it's a slow attack, but they're after me, I'm sure of it. As I peek out of the tent first thing, I see a mass of jet black shiny slimy man-eating yuckiness, all moving relentlessly towards the tent from all directions like the worst of the cybermen from Doctor Who. And then I see one inside my cup, another on my spoon (retch!) and two more on the sides of my saucepan. It's definitely an attack.

Managing to outflank the invaders, I clear the area and peace is restored. I know I should love the slug as I do the midge, but I'm struggling.

I take a walk up over the rough ground behind the tent to get a look northwards. This is a wild place. The vegetation is not close-cropped by deer, and although I can see evidence of them, they're not impacting on the grasses and trees as they do elsewhere. Deep tussocky humps of grasses and sedges, streams underfoot beneath the green (heard but not seen, just the gurgle of water giving them away), brackens, heathers, bog myrtle and trees, yes trees. Oaks indeed. Most are small, close to the ground and/or cropped by hungry mouths, but many too have clearly got away, and on the steeper slopes or within the hard-to-reach rocky bounds several species are thriving.

And so are the ticks! I'm wearing shorts as normal, and am picking up ticks all the time. I don't mind that so long as I brush or pick them off within the hour. Up on the hill, every brush with the bracken is a brush with the ticks.

Compared to the midge, the tick numbers in Scotland are low, but still plentiful enough to find you just about anywhere in the Highlands. They too can excite a kind of horror, and perhaps they too have a part to play in protecting us from the Benidorm effect. Bearers of Lyme disease and even encephalitis, these tiny beasties will find a soft and secret nook, often around the back of your knees or up around your groin, and bury their heads in, sucking your blood. The baby ones are this size – · – and difficult to spot, while the adults crawl about you like flat little spiders, and you don't feel any of them until they're buried, bums in the air, in your skin. To these guys we're just one big blood meal, and while perhaps not as annoying as the midge, they can still scare the proverbial out of second homeowners. Should we love the tick as much as the midge? Absolutely!

/ / / / /

At the top of the hill the most extraordinary views open up: west to Eigg and Rum, north to Arisaig, south to Ardnamurchan and east to the Moidart Hills and Glenfinnan beyond. Blue sky, green hills, open seas, wind, sun, wide-open spaces. I breathe it all in.

Decision made; I'll take a risk and head out into the messy sea and give it a go. Even from high up, where the perspective always flattens lumpy water, it looks lumpy. But on looking more carefully I see that most of the white breakers are in a line from the island channel where wind and tide are butting up against each other. I think maybe I could sneak round the inside of this wind tunnel and use the relative shelter of low tide to tuck myself away too.

An hour later, feeling nervous but reminding myself that I can always turn back if necessary, I set off. A kilometre to warm up, paddling towards what looks like a wall of breaking waves in the gap, and then I'm there. The sea quickly rises, the noise of the waves increases and the intimidation begins. I began to talk to myself, to reassure myself that I can do this stuff, and I paddle into the waves deliberately slowly. Testing, thinking, assessing, letting my body adjust to the movement, the uncertainty. I check in with Wilson too, and he seems happy; time to commit.

The next hour is another of this trip's more intense paddling moments. A sea on the beam that jumbles and breaks, a blustery noisy wind, a coastline without get-outs, and a long fetch to windward, forecasting still more wind to come. All is movement and light and edgy, but kind of wonderful too.

After I break free of the west-facing peninsula, my plan is to turn eastwards and ride the waves on the stern. That will keep me close to the shore should I capsize – but will also take me several kilometres inland, only to have to haul back out just as far against the wind once I reach the north shore. So I change my plans – right there, in the churning seas – and hold my line northwards, across the Loch nam Uamh. It's only a 3-kilometre crossing, and on a calmer day I would not even be thinking about it – but today that 3 kilometres looks a long way and feels very exposed. A capsize would mean a very tricky roll with either an equally difficult re-entry or a long, long swim east before reaching firm ground. Capsizing really is not an option.

Out there in the middle, right on the edge of my abilities and alone is an experience I hope I won't forget. Looming waves, sun bouncing off the sea, boat riding the moving water beautifully, engaging with the yoga breath, feeling 'out there', immersed in the natural world, and at some point even starting to relax. These are the moments that are the real treasure in this journey, an intense sensory input that is hard to put to paper.

As the Arisaig shore draws close, the offshore islands began to take some of the rougher waves off the water and then there is only the wind and a short chop to deal with. I'm wet from the crossing and tired – and, to be honest, relieved to be out of the swell. I turn westwards, into the several kilometres of calmer water that lie ahead, albeit with all with wind on the nose, and I'm glad, an hour or so later, to land – on another exquisite white beach – at the western end of the peninsula.

Ardnamurchan south at last

Battling otters off Arisaig

Sunset over Skye from Uags

The usual graft of lumping kit and boat up above the tide done with, I bask in a breezy late afternoon sun, swim, dry my wet kit and take a walk out to the point. Small islands in the stream, bouncing light, wind and waves onshore, shell sand beach, a natural quartz dike squirming between the granite, Ardnamurchan far to the south and Arisaig and Mallaig round the corner to the north. Yes, despite the weather, this journey of mine is alive and well.

- Route: west Ardnish Peninsula to west Arisaig Peninsula
- Distance: 13 km
- Weather: sunshine and wind
- Wind: Amber. Force 4 to 5
- Sea state: lumpy, wind-blown, at my limits; in the afternoon, calm with wind on the nose
- Events: the walk, nosing into the gap, the crossing, the camping spot, swimming, sunny day
- Camping: to die for.

Saturday 16 July, Day 44: West end of Arisaig to Isle Ornsay on Skye, including ferry from Mallaig to Armadale, 27 km. Wind: Green/Amber.

An early start again to catch the tide flowing north. I'll need energy today, so I've breakfasted well, with porridge flavoured with the dried figs purchased from Eriskay a week or so ago – porridge will never be the same without them. Paddle out round the west end of the Arisaig Peninsula on a calm sea and chase the north-flowing tide to get inside the Arisaig skerries and north to Mallaig before the wind comes at lunchtime. A cooler, cloudier day today with an occasional light wetting, but it is all good. As I come round the point, two otters are battling it out, squealing and scratching at each other, oblivious of my presence. Small white beaches at Keppoch and beyond – mostly empty when I started coming here back in the 1980s, now feeling somewhat urbanised, with a score of white boxy campervans lined up above the shore. Stop to stretch just south of Morar, and chat with a young couple walking their dog. Neil is a tourism blogger living in Arisaig and uses the wonders of IT to make his living. He is an example of the modernising economy coming to the Highlands, and his 'Travels with a kilt' blog is an intelligent and thoughtful contribution to how we think about ourselves here in Scotland and what is our place in the world.

Mid-morning now and the wind has risen, so it is a wet trip into Mallaig, where I lump the kit up to the high street and trolley it all to the ferry. The wind is too high for me to take on the Armadale crossing, and having made contact with my old friends Robert and Pauline, whose boat at the tidal island of Ornsay on Skye is going to be today's destination, I don't want to miss the chance of catching up by unnecessarily waiting out the weather.

A short ferry ride and I'm in Armadale, carrying kit back down to the water and paddling north up the Sound of Sleat in a grey sea. Ten kilometres later and there are Robert and Pauline pushing their small tender down the muddy shore at Ornsay on what must be the lowest tide this year so far. On the yacht, with my kayak bobbing astern on two ropes, we catch up and reminisce. Our two families go back a long way; Robert and I have worked and kayaked together, and our kids grew up alongside each other too. Pauline cooks steak, we drink whisky, laugh lots and sleep soundly, the boat gently rocking with the sound of the rigging tinkling against the mast all night.

Tomorrow I need to get through the narrows at Kyle Rhea for the second time this trip, and will then turn right after the Skye Bridge and begin the final journey north towards the finish. There is a long way to go yet, with several major headlands to take on, but as I'm well over the halfway mark now the journey has taken on its own shape, and I face north with a quiet satisfaction at the miles now under the keel. Gourock feels a long way away not only in terms of physical distance but also mental space. Back then all was nerves and 'what ifs', while now I'm more relaxed, settled to nature's rhythms, accepting of change, bedded in.

- Route: West Arisaig Peninsula to Isle Ornsay on Skye (including ferry to Armadale)
- Distance: 27 km
- Weather: calm then breezy; grey
- Wind: Green/Amber. Calm, then Force 3 to 4

- Sea state: calm, then lumpy on the stern
- Events: otters, Neil the blogger, rough ferry ride, meeting Robert & Pauline, steak!
- Camping: slept on the boat.

Sunday 17 July, Day 45: Isle Ornsay to Uags (southern Applecross Peninsula), 33 km. Wind: Green/Amber.

Quite a day of paddling, but as I am writing this journal two days later, it means I lose some of the immediacy of memory, especially the 'feel' of things. So instead I will just summarise and see how that goes:

Lumpy sea on the stern to Kyle Rhea; planned distance and tide perfectly as Kyle Rhea narrows were calm right on the high tide, which turned just ten minutes after I was through; seals galore; otter close to the narrows; hidden ruined crofts with trees sprouting within; sun comes out; wind rises to Force 3 to 4 and on the stern once under the Skye Bridge; gorgeous green islets heading east towards Plockton; wind becoming tricky, difficult sea for prolonged period so put ashore just before turn to Plockton; outgoing tide leaves my kayak 200 metres from the water; shallow bay stuffed with shells and coral all slowly turning to sand; yoga on the beach; bash through the bracken to check the sea state for the crossing to Applecross Peninsula; wind eases, so back on the water late afternoon and a lovely rolling 3-kilometre paddle over to the far side; wind on the nose up to the turn north; low tide into Uags bothy bay; a Mountain Bothy Association works party is in situ; quiet chat and gentle laughter heard from offshore; camp rather than bothy; invited to join the team for cheese and chat; brutally midgy back at tent; unbelievably beautiful sunset of deep orange and gold.

- Route: Isle Ornsay to Uags (Applecross)
- Distance: 33 km
- Weather: grey then sun; wind all day
- Wind: Green/Amber. Force 2 to 3, then 3 to 4 south-west
- Sea state: on the stern, messy, needs concentration
- Events: crossing to Applecross, MBA works party
- Camping: outside the bothy, midgy, sunset.

7 | Nasty, brutish and short

Days 46 to 51 – Uags to Clashnessie

Monday 18 July, Day 46: Uags to Craig (north-west Torridon), 33 km. Wind: Green/Amber.

Felt like a big day today, tired and a bit sore now. Also very hot at the back end of the day as the wind has dropped. Once the sun has disappeared clegs[10] come out in force, and the midges in their millions. I swim in the pools of the river at Craig. Warm, clear, sun-kissed, smooth brown rocks, diving down, sun glinting through the water beneath.

A list of memories from the day:

Leaving Uags with the gentle chinking of hammers and chisels as the old pointing was replaced; the quiet chat of the two old boys up the scaffold, happy in their work and shrouded in midge hoods; coming ashore at Applecross Inn, meeting Jimmy, ex-drummer for Stiff Little Fingers and enthusiast for the various community initiatives in and around the village; meeting Judy, owner of the inn, who sold me supplies including some wonderful flapjacks (I'd missed the shop 2 miles to the south), friendly, helpful, welcoming; scoffing two of the inn's 5-bacon rolls – one piece of bacon for each of the 100 miles of the North Coast 500 route. I never usually eat bacon, but the wee stall was open across from the inn, the sun was shining, I was starving – and yes, I weakened. A couple of tourists were about, and with my kayak

---

10 horseflies

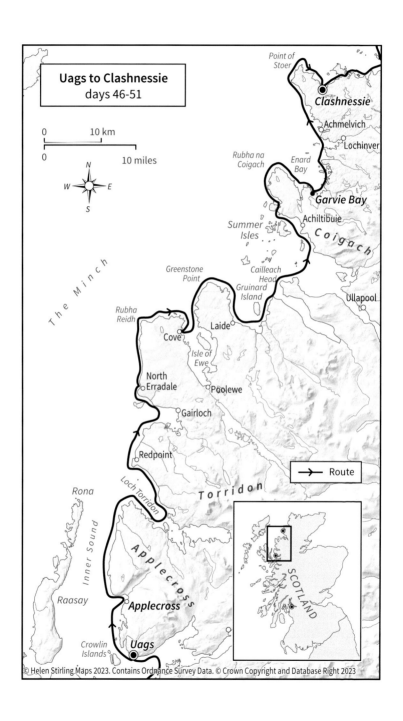

on the shore they showed some interest in where I'd come from and was heading. I remembered to ask about their journeys too; we all have stories to tell after all.

Applecross is another Highland community where there is a lot of fantastic stuff going on. Their community company is busy building affordable homes, running a shop, a petrol station, woodland, allotments, a hydro scheme, other energy-efficiency schemes – and, more recently, a brewery. From the outside in, it's a triumph of community ownership, and I know that beneath the surface successes will be a huge amount of hard graft, most of which will have been voluntary.

As I paddle on northwards, the wind comes on from the south-west and rises from a ripple to a blow. Waves quickly build, creating an exciting paddle in a sun-kissed sea. As I move north the fetch grows and so do the wavy bits, and offshore the sea flashes white with breakers. Rounding headland after headland I have brief moments of respite before the fetch reappears, along with the increasingly bouncy sea. Eventually my nerves give way, so I put ashore on an exposed rocky coast amongst boulders, rest on the wet grassy slope, stretch and recover some energy.

The conditions are not due to change, so after an hour I get stern with myself for prevaricating and head back onto the water. After 30 minutes of balancing on the waves, I find the wind does finally begin to ease and I can relax. Rounding the northern tip of Applecross and seeking out a possible camp spot, I see that all is gnarly rocks and wet lumpy ground. Nowhere to camp here. Pressing on and round and into Loch Torridon then turning left to cross the 4 kilometres of open water to the northern side at Craig. The Torridon Hills, looming blue in the early evening light, provide an immense vista, a vast open space. I know these hills well. and coming amongst them from the sea I feel comfort in that familiarity. But there is also a healthy dose of respect as the high tops can gather serious weather, and I have many memories of tramping about them in all seasons, all weathers, with friends, with my son, on my lonesome and even in a mountain marathon or two. This part of the journey is bringing me back into the parts of the Highlands I know best. The finish, up where my life here, north of Lochinver, began so many years ago, seems not far away. But there is plenty of unknown ground ahead too, not least the series of exposed peninsulas that march north from here: Red Point, Rubha Réidh, Greenstone, Rubha na Coigach and Stoer Point. Any one of these could pin me to the shore or dispense with me out at sea if I misjudge wind and tide – and my own abilities.

From the southern Torridon shore, Craig Bay is just a distant smudge of beige against the darker grey of rocky shores either side. Setting out into the rolling blue sea, I relax and fair glide across the bay. Slowly the distant smudge takes shape, details appear – and then a kilometre out it feels like the crossing is done. It is hot now, and low tide and a smooth bouldery shore makes for an energy-sapping haul of kit and boat up to the grass. There are ruins here, and thick vegetation has reclaimed all but a tiny scrap of flat grass that becomes my camp. Pouring with sweat and with body feeling a bit dead, I try to move slowly and deliberately. The clegs are out in force once again, and I'm bitten pretty well everywhere. My usual good humour with the tiny beasties, including the midge of course, is well tested – and yes, I do on occasion lash out. Sometimes self-preservation takes precedence over principle.

Once settled ashore I brew up to rehydrate, and then take that walk out to the river at the northern end of the bay. The place is gorgeous, and I am here, miles from anywhere, entirely alone. Rocks worn flat and smooth over time, water bubbling between pools, warm sun, warm water even, strip off and swim, fully immersed in this little piece of wildness and laughing out loud at the sheer joy of it.

Back at the tent, I find the wind has dropped and the cloud cover has brought out the midges in force. Zip up, mozzie coils going, and try not to scratch. It's a hot night, and midge city when crawling outside to check the stars. Not restful. But the sea is calm, and there's the promise of progress tomorrow.

- Route: Uags (Applecross) to Craig (Torridon)
- Distance: 33 km
- Weather: sunny
- Wind: Green/Amber. Calm then Force 3 to 4, eased at north end of Applecross
- Sea state: calm to wind-blown sea at stern and port

beam; at one point I come off the water to get rest from
the lump

· Events: kindness of strangers (Applecross folk, Jimmy
and Judy); deciding to go on, even with lumpy sea,
crossing to Torridon, clegs, swim in the river, carrying
kit miles up from low tide

· Camping: bit gnarly, and lucky to find. Gorgeous views,
and a river swim to die for.

Tuesday 19 July, Day 47: Craig (Torridon) to North Erradale (west of Gairloch), 22 km. Wind: Green.

Beautiful day of scattered sunshine, warm breeze from the west-
north-west, and good progress too. Forecast has suggested lots of rain
from late afternoon, so I lay out plans for the day to get ashore and
tented by then.

Out of Craig on a flat calm and up to Red Point where a sea
eagle, large and scraggy-looking and flapping hard to get airborne,
heaves itself up from the rocks just above sea level, to be immediately
mobbed by terns and gulls. These birds pay the price for being an apex
predator. Even though they don't predate on terns, I suspect that in the
terns' minds they remain, like any raptor or corvid, a potential threat;
clearly, better to attack and harry first rather than wait, and risk being
a victim.

The beach on the south side of Red Point is almost pink with
stone-crushed sand, and the old abandoned fishing station looks out
on an unusual piece of geology. A long low line of basalt, blocky and
broken by wind and weather, gives clues to ancient lava flows and now
protects the station from the worst of the westerly seas.

Round the point, crossing the north-facing bay a kilometre or
so out ... and as if out of nowhere dolphins erupt from a blue sea.
What a treat! Moving fast, a pod of around eight common dolphins,
porpoising with intent, perhaps chasing fish. Flanks glistening green,
grey and white, light flashing off dorsal fins, taut muscle-bound
athleticism. And then they're all around me. I'm paddling hard and
laughing with the joy of it, this visitation. A few of them leap clear of

the water and twist to splash down on their backs and sides, they criss-cross each other, jinking this way and that – so *fast!* There is an intense exuberance about these animals. They come close, within 5 metres or so, circle about – maybe checking me out, maybe not – and keep to their northerly direction. Then, a couple of hundred metres ahead, they just stop, fins moving slowly, randomly, on the surface. Have they caught something? Are they in kill mode, or eating perhaps? As I draw near, they kick up a gear and are off again, fast, energetic, purposeful, free.

I attach these words to these animals but, as explored earlier in the trip, I cannot really know what they're thinking or what they're feeling. Sometimes I think you can make a good guess, though. The two otters I spent time watching at the mouth of Arisaig Bay days ago were clearly not pals. There was squealing and teeth and face-offs and biting and chasing. Seemingly oblivious of my presence, was this a male otter setting to mate or perhaps chasing off a rival? I don't have answers, but it certainly didn't look anything like the play behaviour I've often witnessed elsewhere. Play, for otters, also has a chasing element, but there is a fluid, rolling lightness about their interaction then. These two in Arisaig, they were at each other's throats.

And the dolphins? Do they play? What is riding the bow waves of boats or surf if it's not play? But the animals I saw today appeared to be moving with purpose, moving together. And then the change, the stop, the quiet fins at the surface – and then away again, this time with more jumping and twisting. Perhaps a hunt and a kill, some form of social messaging within the pod, or maybe just sheer exuberance, fit animals bouncing about and around each other. Whatever it was for them, for me it was pure joy.

I pull into Opinan Beach with its deep orange sandstone sand, and swim in clear waters under a warm sun and a light north-westerly breeze. While it's initially empty, a few folks emerge from the road end, and some come inquisitively to the boat. While feeding and stretching I meet Steve, an academic ecologist from Glasgow University and expert on all things cockles. Sometimes you meet someone who you'd love to spend more time with, and Steve is one for sure: Glasgow-born, clearly deeply knowledgeable of his subject and of the broader education piece, and dedicated to doing good by the environment. Cockles, it turns out, can tell us much about the changing environment, not least around the acidification of our seas, which affects their ability to form their calcium shells.

With time marching towards the predicted afternoon rains, I take my leave and cross the 4 kilometres of open water to the Gairloch Peninsula. As I slowly haul in towards Longa Isle and make my way north to North Erradale, I have a decision to make. Ahead lies a 13-kilometre west-facing peninsula out to the Rubha Réidh lighthouse, and then a 6-kilometre ride across the top before shelter. There are few get-outs and even fewer places to camp. It is an imposing prospect, especially with the wind due to rise from the south-west slap bang onto this coastline. As I begin to turn north the heavens open and rain pelts down. While I love the sight of the drops on the water as I paddle along my mind is on getting ashore and keeping my stuff dry while getting the tent up. With the rain comes the rising wind so, decision made, I put in on the difficult rocky shore at North Erradale. A more slippery, gnarly landing I can hardly imagine, but beggars can't be choosers and at least I'm safe. An hour later I have everything up on the grass, tent pitched and am in, out of the rain at last.

Tomorrow the wind will go up a notch. Tomorrow, too, I'll need to let my friend and neighbour Ken know where I'm going to be on Thursday, as he's going to join me for a couple of days. Another friend, Rog, is also joining me for the final week of the journey. Both say they're coming to help me re-socialise into society after too long on my own. It's a joke, I think.

I'm looking forward to having a bit of company in these final days, but am aware too that the tone of the trip changes when it's no longer just me on my lonesome. The solitude and exposure of being alone comes to an end, and I'll need to get used to making conversation again and sharing decision making. I have been solo for seven weeks now, and have become really quite self-reliant. I have a system for the tent and for the cooking, and I alone decide when, how far and how hard to paddle. That all changes soon, so I need to be ready for that. Having some company will be strange but lovely, and two more able and understanding pals like Ken and Rog I cannot imagine.

I will miss this solo travelling, though. Alone but not lonely. Exposed and often on edge, but not, apart from a couple of exceptions, terrified.

Self-reliant, utterly, but accepting of the kindness of strangers, and on two occasions my lovely wife's support with the vehicle. Travelling alone, I've had time to focus on a clear aim, to be thoughtful of what I'm seeing and doing and to take time to think, write and absorb the natural world of the Highlands as they are and not as I'd like them to be. And then there is the resilience bit. Being alone has helped me build on that. I've learned to accept the weather, to be a bit brave, to take some risks and to allow the joy of the natural world to wash over me. Solitude and self-reliance have been the tonic, I think.

I will remember the people I've met and the work they do to create communities up and down our coast. I will remember the welcomes I've received; the fish from the Lochboisdale fisherman; the ladies in the community shop at Eriskay who couldn't do enough to help; Neil, the young blogger at Arisaig, engaging and educating us all on the social and political issues hereabouts; the wonderful Swiss ladies who helped with my Harris portage; Andy and Wei at the first portage at the other Tarbert; my fishermen friends Iain and Ben for the meal on Harris; Robert and Pauline on the boat for the steak and the bunk; the wrasse fishermen on North Uist sharing their thoughts on the ethics of fish farming; young Jane biking through the Highlands and taking time out to tramp to Uisinis; Judy, the owner of the Applecross Inn, looking out flapjacks to die for; the Kiwi owners of the Salen shop for taking me in out of the rain and the midges; the Calmac folk, always ready with a friendly bit of banter; the 'good lucks' from the occasional yachty; and of course the love and interest shown by my friends in support of my journey, especially Rog who himself had ventured forth on a similar trip some 25 or so years earlier.

This journeying, this full immersion in all that the land and seascape of the Highlands has to offer, has made me happy. Yes, I'm happy here with solitude as a constant. It's a different solitude from that experienced in remote southern Patagonia. There Robert and I journeyed in places where there were no people for many miles and where we met no one for days, a truly wild place. But there we did at least have each other to lean on when times were tough. Here in the Highlands, while there have been more people about, the meetings are occasional and fleeting and, whether on the water or in my tent, I'm always alone and must look after my own physical and mental state.

And here, too, there's still a relative wildness, spaces empty of people where animals go about their own business and where the weather reigns supreme. And there is that wonderful full immersion and a freedom to move amongst it all, independent of rules and others' preferences. I love that freedom. And taking on this journey all alone has enhanced the experience and encouraged a level of introspection and external observation that has not been available to me for a long time.

- Route: Craig to North Erradale
- Distance: 22 km
- Weather: calm, light north-north-west, sunny spells, warm – paddle in t-shirt, shock horror!
- Wind: Green. Force 2 to 3
- Sea state: calm seas, then rising with the wind mid afternoon
- Events: sea eagle, dolphins, meeting Steve
- Camping: in amongst the sheep and their poo above a slippery, rocky beach and facing into the coming weather.

Wednesday 20 July, Day 48: Weatherbound at North Erradale, 0 km. Wind: Amber.

A threatening sea greets the morning. The forecast was only Force 3 to 5, but where there is 4 there are waves, and because the wind is blowing slap bang onto this rocky coastline the sea state is messy. The energy from the waves piles up against the steep rocky shore and then bounces back to meet those incoming; a zone of clapotis a few hundred metres wide all the way up the coast is the result. As the angle or geology of the coastline changes, my paddler's eye view, which is around a metre above waterline, gives me scant warning of such a sea ahead. But once you're in it, you know it. Clapotis in miniature looks a bit like thousands of falling raindrops, each leaving its own peak as it hits the surface and bounces water up skywards. Imagine those drops sized up to the bulk of a sit-upon mower, then times that by thousands, and you have yourself a severe case of clapotis right the way up to the headland. When wind meets tide flowing in the opposite direction the water sits up yet another notch; any opposing forces in liquid form lift the water up into a wave which then quickly collapses under its own weight. Up and down, peaks and troughs, confused, no pattern, 'jubbly'. That is clapotis.

So this coastline in this wind and sea is already a confused 10-kilometre line of clapotis 100 metres or so out from the shore. Further out, the waves will be slightly more regular, but still it is messy everywhere. I decide to wait, to climb a hill to find a signal to get an updated forecast. In the end I have to climb two hills and then, tucked down in a shallow hollow, sheltering from the rain, I use the new information to form a basic plan. Wind today is due to rise to 4 to 5 (I was right to delay) and then back down to 3 to 4 tomorrow. That's when I'll go. Not ideal, but it is at least a plan.

The rest of the day is spent repairing the boat, which has taken a hammering over the past couple of weeks. I need a few hours of dry to perform decent repairs but only get two, so have to make do with Gorilla tape to hold the worst of the sea out.

Out in the bay, all is movement and light and liquid jumble. A view that is never still. And the sound too – the wind over water and the waves expelling their energy into the rocks on the shore – is ever present. This beach is a boulder beach. Huge 2-metre-square blocks at low tide merge to large rounded boulders up to 1 metre diameter and at the top football-sized chunks, all rounded by the tumble of tide over time. We walk these places, but how often do we take time to wonder at the raw power and the time it has taken to make them? The relentless coming and going of tides, the energy-rich wind-driven waves, the temperature variance, the rain, all endlessly, massively, grinding away at the rock. The result of all that, of millions of years of erosion, of ocean and geological heave, is this wonder, this massive pile of beautifully smooth rounded stones, each with its own colour and markings, its own geological secrets for those with the eyes to see. It is the uniformity of these boulders that gives mc cause for wonder, too, each stone so similar in size to its neighbours and sorted by storm surges into stratified lines up the shore. A raw power hard to fathom. Only at the bottom of the spring tide does that uniformity break down. There the sea covers the rocks most of the year, and the

Bottling it, Applecross peninsula

Change is coming

Rubha Reidh point

sculpting power from the surface waves is lost.

At the top, above the tides, there is a mixture of boulders and pebbles, the latter hand-sized or smaller. These are blown here, I think, or thrown up, bounced off other rocks, when the sea and wind is at its most extreme. In Shetland a year ago, I camped in a spot about 15 metres up from the high tide. All around me on the grass were broken shards of rock, splinters thrown up from the shattered cliff face below. Standing in that same spot in such a storm would have left me shredded by a natural shrapnel. I have stood on clifftops facing into storm force winds, but have always been too high to notice the rock storm that must have been filling the air not far below. Just as well!

The gannets are diving. Even in this wind. It seems especially on wind-driven seas does this wondrous bird seem to thrive. Today, with bird flu on the rise, there are questions around the 'thriving' piece, but out there in today's chop the bird is doing what it evolved to do – to dive, to rise and soar, to flick and turn a flash of black and white, and to dive again. Poet Norman MacCaig describes the moment the gannet hits the water as 'tombstoned by a fountain'. I love that image, and while the birds hunting in this bay are too far away for me to see the fountains, his words conjure the image as I watch.

How do they see anything from that height, moving at that speed with seas that messy? Are they diving for an individual fish, or is it just an impression they spot from up there – a change in colour, a moving shape perhaps? Are they diving blind, in hope, on instinct? No diving, no fish – no fish, no gannet. Perhaps it's just a play of the law of averages. A shapeshift below the surface, and a deep instinct to fold and dive just in case. Once under water of course, they then need to be able to see their prey, to convert air-seeing to water-seeing. Do their brains account for the diffraction, or is it just about getting close enough to lunge and snap? Others have written eloquently on the habits of our seabirds, and while we can pin them down to a degree, there are still things they do that are beyond us to explain. The scientist and the poet have equal realm here.

These birds feel utterly wild to me, and the distance between their lived experience and mine blows me away. Their endless quest for food, the plunging, the soaring flight over vast distance, the ability to inhabit three dimensions with apparent ease, all these are wonders.

These generalists, suggests Adam Nicholson in his book *The Seabird's Cry*, will be what are left when the specialists such as the puffin and guillemot are no more – for that, appalling as it sounds, could happen. Out there on the open ocean, beyond our ken, the trials of life are very tough and very real. As climate change speeds up, we are already seeing that warming waters are redistributing prey species in the North Atlantic. When the plankton, and the small fish such as sand eels and herring that feed on them, move north, the short wings of the puffin and the other specialists with whom they share their summer breeding grounds may not be sufficient to get out to the new feeding grounds and back to their chicks. The result? Abandoned nests, dead chicks, a lost generation. And then what? These lost lives are uncomfortable to think about, but these are the climate change realities we have to accept.

What we do not have to accept, however, is our own human-induced negative impact on that environment. We observe the animals around us to understand what they need to survive and to thrive, but we also need to observe the environment surrounding them to know if their ecological needs are being met. If we observe a decline in either the species or the environment, then the next question we need to ask ourselves, as whale scientist Rebecca Giggs points out, is, 'Is it because of us, and what are we doing about it?'

Because it is in the doing that change occurs.

And Giggs is right to ask the question, because it *is* because of us. Not just us – but we are and always have been part of the change that occurs on land and sea, and increasingly now in the air. Our impact is not new. We are not, and never have been, somehow separate from 'natural' nature. Humans, and the hominid species before us, are as much 'natural' as any other animal. We have been around in our various forms for over 3 million years, 1.5 million of which we have hunted with ever more efficient tools and with fire. And in that time we have competed for resources with other animals. Where we have outcompeted them, those other animals have gone hungry, been reduced in number or even become extinct. When ancient megafauna are removed from the plains of the Americas, fires are set in the Australian bush, whales removed from our oceans, giant moas hunted to extinction in New Zealand, plants selected, soils turned, lions pushed off their kill by hominid scavengers – when these things happen over one and a half million years or more, depending on your archaeological preference, nature is affected, reconstructed.

Perhaps because we are now so removed from the so-called 'wild' world and all its messy realities, many of us tend to think of humans as somehow outside nature. Indeed, if you look up 'nature' in some of the major dictionaries you will find that the definition excludes humans. This narrow concept is seriously misleading, because we humans, clearly animals, are not outside nature; we cannot but be an intimate part of it. Yet we have constructed ideas such as 'wilderness' and 'pristine natural environments' that cement our sense of separation from nature. What we like to call 'wilderness' is an intellectual construct jam-packed with assumptions, including assumptions that the past was some sort of pristine natural idyll and that early humans were better at managing the environment than we are today.

At one level that is probably correct. There were fewer of us, our technology was simpler and we lived shorter lives. So our impact on the world as a whole was comparatively low. But we were not guiltless then, and we're not guiltless today. Burning undergrowth, chopping trees, killing animals and fish, selecting plants, dropping plastic, laying longlines, turning soils, gathering shellfish – all these activities, some recent, others over millennia, have gone to create the environment as we experience it today. Humans, just like any other species, are one part of nature, and if we – and in particular our decision makers – can only realise that at some meaningfully fundamental level, then we might stand a chance of bringing environmental science and political and economic decision making in line.

Out here on the Atlantic edge, exposed as I am to wind, waves, the natural and crofting landscape and to the literally millions of animals all around me (think cockles, jellyfish, sand eels, birds, insects and mammals too), I am doing my best to observe, to learn, to clarify my own value systems and ultimately to decide what actions I am going to take to make my own positive contribution.

For what is this journey if it's not about taking action on the back of such learning and observation? We journey out to journey in and all that, and learning is all very well – but what use is it if, like a dusty old PhD, it then sits on a shelf doing nothing? When we act, that is when change occurs. Where farmed animals and their environmental impact are concerned I have indeed been active in encouraging some hard-nosed changes on the ground, and will continue to do so. I have worked, too, to promote native, and to control non-native, flora – but where the wild animals are concerned, I know I could do more.

I have learned that I can do so closer to home than I might have expected, for making changes close to home can counter negative impacts both local and global. You don't have to travel to the other side of the world to protect turtles from plastic pollution or stand in front of a forestry machine to stop deforestation in the Amazon. If we all did that then our travel carbon footprint would be bonkers. But what we *can* do is change our behaviour here at home, and by doing so positively affect a wider environment. 'Think global, act local' works, and it's empowering.

In my own backyard the plastic flotsam and jetsam I'm seeing on the remote shores across the Highlands stem mostly from our everyday purchases and from local sea fishing and farming activities. Further afield, the burning and logging of the Amazon making way for soy is driven by the billions of unquestioning purchases of processed foods or intensively farmed meat. By not asking questions about how our food is produced we are part of the problem. And on a global climate level, the dreadful floods in Pakistan in 2022 were the result, it is argued, of a wider atmospheric warming that melts mountain ice. Such warming not only melts ice but carries moisture into the heated air which, when saturated, dumps it as rain, thus further swelling rivers and flooding lands. If we drive a car, fly or buy carbon-heavy products then we have played a part in that too.

What we do locally matters globally.

We know these things – but do we care enough to act? 'Knowing' and 'caring' describe diverse ways of being. Knowing is about understanding things; it's intellectual, rational and scientific. Caring, on the other hand, is a feeling, steeped in the emotions arising around empathy, kindness and concern.

Knowing, on its own, seems worryingly ineffective at exciting action at a political or legal level; you need only look at the graph of ever-rising atmospheric carbon levels during the 26 years of UN climate change conferences known as COP (Conference of the

Parties) to see the gap between knowing and doing. To get action we need some emotional motivation too, more care and empathy for the environment that supports us. Lots of people in the developed world do care about the environment and there are many wonderful positive actions taking place. But as a collective group, as a society, and despite all our knowing, it seems to be difficult for us to bed such caring into political or legal action. Our politics, where big-picture, long-term environmental care is concerned, lags way behind the science, and my suspicion is that it is because the care element, the deep emotional attachment to our environment, is still missing. Perhaps it needs the global warming piece to have a swift and powerful effect directly on our decision makers. It's worked before: the Great Stink from the Thames in 1850s London and the same city's 1950s smog left MPs gagging and gasping respectively as they sat in Westminster. The results? A sewage system to clean the river and a Clean Air act to, well, clean the air. Perhaps the 2023 Canadian forest fires shrouding the skies of New York might cause politicians there to act now their suits are reeking of smoke.

There is plenty of research suggesting that our more authoritarian religious systems and our science have until recently been almost blind to ecological abuse and the negative animal welfare that results. For many years, especially here in the western developed world, we have objectified our environment and the animals within it as subservient to our material needs. Too casual or selective readings of Christian religious texts have created a belief system that casts humans not just as separate from nature but *above* it. That theme has served humans well, especially in western cultures, for the creation of material wealth. But that very same wealth has made us spiritually and ecologically poor – and ironically enough, materially poorer too, in the longer term.

But look further afield, to Buddhism, to Shintō, to the beliefs of those who we call 'native peoples', and you see humans, animals, plants and spiritual beings merging in a constant interplay of social and political interaction. No one species stands alone; all are mutually dependent in a symbiotic relationship of one sort or another. The Native American worship of fish and other fauna, the Australian Aboriginal sacredness of place, the Amazonian Indian reverence for the forest, all demonstrate a different sort of knowing, and – most importantly I think – a reverence and corresponding care for the environment that supports them.

In the developed world we have lost that inclusive way of thinking, and our natural world, with some exceptions, is not yet bedded into our value systems or our ways of being. We can learn much from those that do – those who, with a little reverence, a different form of knowing, demonstrate a wisdom that we are sadly lacking.

For reverence is a powerful motivator, and I would argue that without it we are unlikely to change our behaviour.

Changing our behaviour itself can feel like a sacrifice. But giving up that second holiday abroad, that long-distance flight, the second home, the unprovenanced steak, the uncertified fish supper; and then taking actions such as insulating our homes, reducing our waste, investing responsibly and making it clear what we're doing; these are the sort of things that can have far-reaching positive effects. We should go further, too, and advocate for substantive political and legal change and, if we don't want to do it directly then it's not difficult to use our vote, or to support a non-governmental organisation or two. Our individual power, after all, lies within us not just as consumers but also as citizens.

Some suggest that individual action is like a single drop in the ocean, invisible and insignificant. I agree that when working alone our individual efforts can seem small. But I contest that they're insignificant. Big change, sustainable change, first appears not from big government but from the minds – and the words and the actions – of individuals. Think scientific and industrial revolution, think abolition of slavery, civil rights, feminism, the ban on DDT and chlorofluorocarbons – indeed, the green movement itself – all were changes that emanated initially from educated and passionate individual minds. And it's when those individual minds band together that exciting things happen. Opinions and new thinking gather momentum and bearing in mind that to create the tipping point for realising change, as the revisionist Malcolm Gladwell suggests, we need only around 20 per cent of individuals within a given population, it is clear that the heated and hopefully well-informed discussions in the pub, those so-called drops in the ocean, do really matter.

Of course, there is also enormous potential power in government to encourage behavioural change, but the politics are so often complex and combative that change seems too slow in coming. Our species, for all its self-vaunted intelligence, is remarkably short-termist, and appears less rather than more likely to take the necessary steps to survive, even as that survival becomes increasingly threatened. We need more individual discussion around these issues – more emotion, more feeling, and certainly more reverence. Perhaps an old film will illustrate my point, and a Scottish-based film at that: *Loch Ness*, starring Ted Danson and Joely Richardson. Ted arrives on the shores of Loch Ness to disprove the existence of the monster and promptly falls in love with single mum Joely Richardson. Joely's little girl knows things about the monster that the adults don't. In amongst the somewhat schmaltzy storyline is a golden moment when Ted Danson's character, Mr Dempsey, is tucking the little girl into bed and he's saying that after all his fruitless searching for the mythical beast, only if he sees it will he then believe it. 'No, *no*, Mr Dempsey', says the little girl. 'First you have to believe it and *then* you can see it.'

It's about faith, isn't it? Faith that there are aspects of our world that may be beyond most of our ken but which are fundamental to our existence. Beyond our TV screens and our daily horizon, there is a struggling ecological system, a declining health of our soils and our flora and fauna, and possibly indeed even a mythical beast – why not? All are difficult to understand – and gathered as we are into cities, invisible to most of us. So let's have a little faith. Faith in the science that surrounds climate change; faith that the environment-positive 'sacrifices' we make in our daily lives are not actually sacrifices but investments; and faith that where we lead others will follow.

For me, such faith is born out of reverence. Reverence for the environmental system of which I am one tiny symbiotic part. And I am not suggesting reverence as an overly simple 'worship' here, for that could, as discussed earlier, lead to separation of human and nature. What we need is the opposite. We need to see ourselves as included within nature and not separated from it, and thus show some humility and some restraint. With reverence, heartfelt and backed up by good science, we can make decisions on our activities that more carefully consider the mutual impact that humans and environment have upon each other. The endless quest for economic growth, generally devoid of robust enough environmental impact assessment, is killing us, because it is almost always a one-way transaction. The environment gives and we take. Our every action, whether it be as individual, as business or as state, is both influencing and influenced by our environment. If enough of us can realise that, then we stand a chance of bedding in environmental impact decisions as cultural, spiritual and legal values that will colour the political and economic actions we take …

Back on my rocky shore, the rain spattering hard down on the tent, I know my journey out has been everything I hoped for: challenging, inspiring, at times frustrating, weathered, often fearful, always engaging and filled with that most precious of commodities, time to think. Thinking itself is a kind of doing, of course, and perhaps the real gain is indeed the journey within. The result? Well, so far it is a change of perspective, feeling small and humbled in the vastness of nature, being in awe, re-aligning what is important – and, most importantly, starting to tease out what actions to take post trip.

- Route: weatherbound at North Erradale
- Distance: 0 km
- Weather: windy, wet in parts
- Wind: Amber. Force 4 to 5 south-westerly, and going north-westerly tonight
- Sea state: messy
- Events: nothing of special note
- Camping: as yesterday.

Thursday 21 July, Day 49: North Erradale Bay to Gruinard Island, 45 km. Wind: Green/Amber.

A long exhausting day, alone out on the outside edge of two of the most exposed headlands the Highlands have to offer. The winds, as forecast at least, looks acceptable, at most 2 to 3 with occasional lifts to Force 4, all of which suggest I should be okay on the water. However, the direction is north-westerly, a continuation of the westerlies that have blown hard now for a couple of days and which I know are blowing straight on to the exposed coastline ahead.

I set forth on a grey day with hope that the lighter wind will have taken the edge of the sea. Out from Erradale it looks okay at first. However, by the time I get north of Melvaig, the final tiny settlement on this remote coast, the waves have built and it has started to get messy. The underlying swell is not too bad, but the residual energy from the waves and wind over the past days is still in the water; waves a good bit more than twice the height of my profile in my boat keep coming at my port bow, and each wave is splitting into others as the energy bouncing back from the shore meets the energy still coming in. Confusing! Intimidating!

So, rapidly filling up with self-doubt, I have some self-talking to do. I swallow back the fear and the instinct to run for shelter, have a few stern words with myself and make up my mind to get on with it. I have this, surely by now I have this. What 'this' is I am not yet to know for sure, for the 15 kilometres or so ahead involve paddling through an increasing height and complexity of waves. Now well and truly committed, I hold any panic at bay with a steady stream of, 'you got this', 'hold your technique', 'paddle it', 'don't freeze', 'let the boat do its thing', 'breathe!' I lean on the sea, reaching over the waves as they surge towards me; pull hard, lean again, and repeat. Every instinct says 'get close to the shore' but that would put me in the clapotis zone and the sea will just get worse. It's better to stay out, sometimes for up to a kilometre, which out there on my own on the edge of the world feels like a very long way.

Digging deep mentally and physically, stroking hard, yoga breathing, swallowing the angst ... and slowly the headlands ahead come level then drop astern. Then, after I don't know how long, the steep ground begins to lower, then quickly drops away. The point; I've reached the point. This of course is where wind and tide meet and where the unwary paddler can come horribly unstuck. Sure enough, at the turn everything rises another notch, the waves shorten and steepen, my nerves frazzle, and all I can do, boat leaping about on the short steep chop, is paddle hard through the tidal mess. So focused am I on the oncoming waves that Rubha Réidh lighthouse appears only in my peripheral vision, and I target an outlying rock, behind which I capture a few minutes of respite from the churning water beyond. Tucked in here, with the lighthouse above and the noise of the breaking sea all around, and holding position in the swelling water between the rock and the shore, I feel as small and exposed as I have ever been on this journey. Ahead, the north-facing coast of this peninsula stretches east for another 6 kilometres, and any hope of a let-up at the turn is banished. The sea rolls and pitches, the shore is white with breakers and untouchable, and safe water is still some time away. 'Keep your head', 'hold your technique', 'paddle it', 'breathe!'

I make it, of course, 6 kilometres of unpleasantness later. It only takes one wave to knock you over, and alone in that sea I would struggle to self-rescue. Again, capsizing is not an option. The ferocity of this exposed part of the coast is made real by the ships that have foundered here. I'm aiming for the gap at the island of Furadh Mòr, where in 1944 tragedy struck as the American Liberty ship USS *William H. Welch* broke in two on the rocks at the cost of 62 lives. For me today, though, the same island is a sanctuary, and as it comes close the sea begins to ease, the waves lengthen, the underlying swell becomes more apparent and the sea moves behind me. Unseen waves, now at my stern, pick up my boat and give me a surge. It is still too big and chossy for me to enjoy a surf, but there is relief as the safety of the gap looms ahead. I let the sea, the energy in the water, pass under me, and it feels like I'm riding a series of shelf-like ridges, each leaving me momentarily balanced on its edge until the rest of it slips by beneath. The nose of my boat first drops into the trough as the wave lifts the stern, then rises high as the same wave passes through. I am a nodding donkey, swiftly tilting my way to calm waters ahead.

Through the gap and into Loch Ewe – and I'm into a different world. The sea calms, there is no inkling of the very different conditions outside the loch, and before too long I'm paddling into the tiny community of Cove at the end of the coast road on the western shore of Loch Ewe. From here in the early 1940s the Arctic convoys set forth to relieve beleaguered Russian ports north and east of Finland. These waters, now quiet, were a bustle of wartime purpose and hulking steel as thousands of sailors went to the rescue of their allies, and many to their deaths in the freezing conditions of the Barents Sea. On a sunny day Loch Ewe is idyllic, but to see it with wartime eyes is to take a glimpse behind the apparent remote and scenic view to a globally connected, and in this case very real, human tragedy.

As I paddle into Cove, all of a sudden the energy goes out of me and my arms feel heavy as lead, a reaction perhaps to the intensity of the past couple of hours. I need food. It's at Cove that I've planned to meet Ken, who is to join me for the couple of days before Rog arrives for the final week.

But Ken is running late so, after drying out and feeding up I decide I've not had enough excitement for one day and set out to round Greenstone Point to meet him at Mellon Udrigle in Gruinard Bay. I can see across the mouth of Loch Ewe towards the point at Slaggan Bay, around 6 kilometres away, and Greenstone Point is another 5 kilometres after that. There is then a 4-kilometre stretch of unprotected coastline along the top before the turn into the shelter of Gruinard Bay. Plenty of exposure, then!

So, nervous but slightly buoyed that I've handled Rubha Réidh on my lonesome, I set off for Greenstone. But as I'm paddling east across Loch Ewe the sea starts to rise and by Slaggan Point I'm back on the same big messy stuff as before. Right back at you!

Two and a half hours from Cove, a very frazzled paddler finally strokes into Mellon Udrigle. Ken and his son Angus are waiting on the beach and I do my best to pretend I'm okay. On the surface I'm weary and wet but cheery enough, I hope – but underneath the façade my nerves have taken a battering. I have that taut jangled feeling you get when you wake from a disturbing dream where you can't get away, can't shake the threat. I really need some calm sea for a while to help me settle.

Angus bids Ken and me farewell and heads home with the car, leaving us to paddle off for a camp spot on the southern tip of Gruinard Island, 7 kilometres away. This is the first time for nearly two months I have had a companion, and we will pick up with Rog tomorrow.

Touching the island shore, Ken and I are in mind of the history of this place. Gruinard was for years known as Anthrax Island. In the 1940s it was used as a testing ground to better understand the effects of the bacterium should it be used against the UK as a biological weapon. As a result, for the next 50 years the contaminated island was out of bounds to the public, and it was only really the pressure put on government by the local Dark Harvest warriors that put this right. It's a great story of derring-do by as yet unnamed academics and local folk who secretly gathered some of the island's soil and delivered it to government property in a couple of parts of England. That galvanised government into action, and the island was eventually cleaned up in 1990 and advertised as contamination-free. Even so, as we put ashore Ken double-checks these factoids before we decide to settle.

A low-tide carry up the steep pebble beach, tent up and, with 45 necky kilometres under my keel, I collapse into an exhausted sleep.

- · Route: North Erradale, Rubha Réidh, Cove, Greenstone, Gruinard Island
- · Distance: 45 km
- · Weather: dry
- · Wind: Green/Amber. Force 2 to 3 and some 4 early on
- · Sea state: the most challenging seas of the trip, the longest and most exposed I have felt to date; wind low but sea high and intimidating, long stretches with no get-outs and the need to stay well out from shore
- · Events: exposed paddling, messy seas, prolonged sections, mind games, meeting Ken
- · Camping: Gruinard Island amongst the midges and the terns.

Friday 22 July, Day 50: Gruinard Island to Garvie Bay, 44 km. Wind: Green/Amber.

Up early as we have only a few hours to get to Isle Ristol, one of the northernmost Summer Isles, to meet up with Rog. The terns are already screaming and tearing about as we head out of Gruinard and straight into a stiff north-easterly blowing hard across Little Loch Broom. The more open crossing to the Summer Isles is a little too white for comfort, so we put ashore at a steep-sided rocky cove close to Cailleach Head to wait for the wind to abate. In calmer weather we would have just cracked on, but in this brisk wind-blown sea, fluky and blowing offshore, and with Ken only on his first full day, we wait it out for an hour then go for it anyway.

Two windy crossings later, noisy lumpy water on the beam, we settle to an easier sea as we near the relatively sheltered shores of the

Resocialising with Ken and Rog – the final few days

Ken, west of Ullapool

Home to the Assynt ranges

Summer Isles, and kayak west between the island of Tanera More and Achiltibuie. Tucked under Ben More Coigach on the mainland, we are kayaking through a beautiful day with a sun-kissed, lazy rolling sea under the keel and the Summer Isles and their outliers creating the perfect paddlers' backdrop. Tanera More was the scene for some of my early reading on Highland life. Frank Fraser Darling described his *Island Years, Island Farm* in gorgeous detail, and I remember being entranced by the sheer adventure of his life out on the edge with his wildlife companions, and the graft in recovering an abandoned croft.

At the Old Dornie jetty opposite Isle Ristol, Rog and Leah are ready and waiting. This will be Leah's last revictualling trip, and she's been a wonder. She's patient and flexible when needed, and without her my journey would likely have faltered early on when the weather pinned me down. She has given me two lifts, each time meaning I've avoided having to wait for the weather to change to allow me to get round Ardnamurchan Point. I've paddled round that most westerly buttress twice before, and it can be a formidable challenge. This time, in this quite exceptionally dreich summer, the wind meant it was just untouchable.

Food rations refilled, Ken, Rog and I set off now as a team of three to take on Rubha na Còigeach Peninsula, the fifth of the six exposed peninsulas describing the route north. I find it very different paddling with friends, and the feeling of exposure, while still real, is dramatically reduced. The sea up to the point is not as gnarly as around the previous two headlands, but it's still a lumpy, confused piece of water, heavy going, and we're all ready for the turn when it comes. This time, as we turn south and east, the sea releases its grip almost immediately, and with wind behind us we enjoy a gentle push southwards to the lovely Garvie Bay.

All over our new horizon are the hills of Assynt, a place I still call home, and a mountainscape I know so very well. Lochinver is not far away now, and back on the shoreline at Gourock, all those weeks ago, I had thought it might be my final destination. Now I can see that we'll get further than that, for sure. How far, exactly? Well, that depends on the weather, as ever.

Camp, swim, massive feed, chat, run from the midges, plan ahead, sleep.

- Route: Gruinard to Garvie Bay, including round Rubha na Còigeach
- Distance: 44 km
- Weather: no rain! Wind, some sunshine
- Wind: Green/Amber. Force 2 to 3 and 3 to 4, with plenty Force 4 gusts
- Sea state: everything in one day!
- Events: climbing the cliff to peer ahead, crossing windy seas in sunshine, meeting Leah and Rog, paddling Rubha na Còigeach, Garvie Beach, swims in sea and river
- Camping: Garvie Bay.

Saturday 23 July, Day 51: Garvie Bay to Clashnessie, including Stoer Point, 37 km. Wind: Green/Amber.

Impressions from the day:

Through Assynt and heading towards northern Sutherland; rocky outcrops; Assynt hills outlined in blue, clouds reshaping the well-known outlines of Stac Pollaidh, Cùl Mòr and Suilven; relatively small hills but looking vast from the sea; crossing wooded bays and the open mouths of the sea lochs – Inverpolly, Inverkirkaig, Lochinver; windy seas at the openings to the east, sunshine, a clean feel to this wind-scoured landscape – and then, dolphins! Exuberant, playful, inquisitive, colourful, free. They're leaping alongside and tilting upside down under the boats, almost within touching distance. Our sociable sea mammal cousins come to say Hi. We laugh aloud at this visitation from the wild.

Achmelvich comes and goes, and the split rock at Clachtoll falls astern too. At the wide white expanse of Stoer Beach, under the ruined broch, Ken peels off to meet Angus, who is driving back up to meet him; Ken's two days are done, and with work responsibilities looming large he needs to get back.

Rog and I paddle on up towards Stoer Point, now chasing a forecast prolonged downpour starting late afternoon. All I can think about is whether this final exposed peninsula will throw the same lumpy sea at us as have the three previous headlands. But the day feels calm and in the end the sea follows suit. We paddle without incident up past the Old Man of Stoer and to the point. There, the easterly wind on the nose changes the tone and it takes effort to grind it out down to Clashnessie, our planned camping spot that night.

Tents up, and five minutes later the heavens open – all night.

- Route: Garvie Bay to Clashnessie, including Stoer Point
- Distance: 37 km
- Weather: breezy but some calm too, dry and warm
- Wind: Green/Amber. Force 3 to 4, felt only at the crossings of the mouths of the lochs, and then the grind up to Clashnessie
- Sea state: rolling on the crossings, calm up to Stoer Point, windy sea back on the nose thereafter
- Events: dolphins
- Camping: Clashnessie bay.

8 | Horizons

Days 52 to 56 – Clashnessie to Kinlochbervie

Sunday 24 July, Day 52: Clashnessie to Glencoul, 22 km. Wind: Green.

A wet morning, very wet. Peek out of the tent on occasion but mostly keep under shelter until mid-morning, when the downpour begins to ease. It's always such a release after the rain. It's the same with the wind, which during this trip has felt more or less incessant. When it eases there is an 'Aahh!', a feeling of release, and an opportunity to make progress once more.

But right now Rog and I are dodging and chasing the weather again. Tonight the rain is to return, and then tomorrow the wind will lift to Force 4 gusting 7. This will put any west coast out of bounds for a day, so we need to find a place to hide. The head of Loch Glencoul will be ideal, tucked away from the open coast as it is, and while it adds several kilometres to our days, rather than being pinned in our tents to some remote west-facing shore we'll probably have options to explore the hills during the gale.

En route to the loch, though, is the old country, the place where Leah and I chose to live, and where we became who we are today. That sounds a bit melodramatic, but it's how we feel. Although we now live in Inverness, it is Sutherland, and Assynt in particular, that still feel like home. Here, in the 1990s, from our off-the-road croft at Kerracher, we worked, had our son Tom, built lifelong friendships,

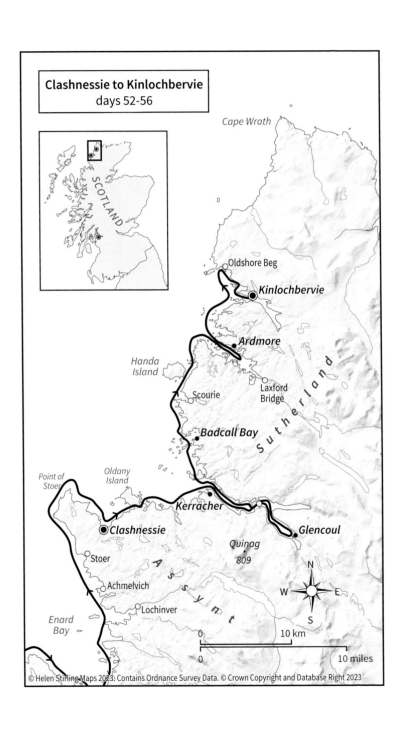

Clashnessie to Kinlochbervie
days 52-56

SCOTLAND

Cape Wrath

Oldshore Beg

Kinlochbervie

Ardmore

Handa
Island

Scourie

Laxford
Bridge

Badcall Bay

S u t h e r l a n d

Point of
Stoer

Oldany
Island

Kerracher

Glencoul

Clashnessie

Stoer

Quinag
809

A s s y n t

Achmelvich

Lochinver

Enard
Bay

N
W   E
S

0   10 km

0   10 miles

engaged with community, took risks including one or two wrong turns, made mistakes, enjoyed some triumphs and had our eyes opened to an unexpectedly gnarly sociopolitical movement right at the heart of community life.

I've kept in touch with the new owners at Kerracher, and while they're not going to be at home as Rog and I paddle past they have let me know they're happy for us to tramp about the place and reminisce. Pulling my kayak up onto the all-too-familiar shore is to take a step back into a life that's now 30 years in the past but which still feels like it was yesterday.

Back in the 1990s, out from Kerracher, I was working as a salmon farmer, or 'grunt' as we were rather cruelly known due to the heavy lifting nature of the work back then. This work kept the money coming in, and with bits and bobs from the sales of a book plus some income from our 'remotest B&B', we made a life for ourselves here at the far western end of Loch a' Chàirn Bhàin.

Kerracher was a place physically apart but intimately connected with the politics of the time, too. With access to it either by walking over rough hill (the landowner would allow no track) or by sea, we lived with the sea and the weather dominant in every day – and we loved it.

Moving in without any telephone and well before mobile phones became a thing, we laid out and dug in 2 kilometres of line by hand, thereby giving us comms with the world. There was no TV, but there was instead the occasional VHS video sent by Grandma (always of a hospital drama – she had been a nurse); and all our food and fuel, everything needed for living, had to be boated in by sea or tramped in over the hill. Once a month Leah would hike out to the car with Tom on her back and head off to the nearest supermarket in Inverness, two hours' drive away. I would then return from the salmon farm early evening, leap in our leaky wooden boat whatever the weather, and head down the loch to Kylesku. We would meet there, and transfer messages, child, coal and any other necessaries into the boat, and head back up what was now, in the winter anyway, a dark loch. Sometimes the journey was a delight of stars and phosphorescence on the water; at other times it was a battle against wind and wet. Back at Kerracher Bay, invariably at low tide of course, everything had to be humped up the rocky beach and into the house and the boat put back on its mooring. It was honest and simple graft; here even the mundane things such as going shopping were an adventure.

Kerracher was an elemental living. On rare occasion, with a northerly gale and a high spring tide, the sea would lap at the bottom step of the tiny front garden. From the small native woodland behind us the birdsong would wake us in summer, and on the croft the broken fences gave the resident bachelor herd of red deer easy feeding on the green. I would delight in walking back from work on dark winter nights and come amongst them, their eyes glowing green in my torchlight. It was the sea, though, that was ever present, always dominating the view. And out in the bay the otters, seals, seabirds and occasional passing fishing boat were a constant backdrop.

The other backdrop to life at Kerracher was political. In 1992 the Assynt Crofters decided to have a go at purchasing the ex-Vestey estate for themselves, and in so doing they revolutionised land ownership in Scotland. The resulting community ownership has since then gradually increased all over Scotland, and especially so here in the Highlands. I was there at that extraordinary meeting in Stoer Village Hall when the decision was made, and while Leah and I don't pretend to have played any more than a background supporting role, when we eventually had to leave Kerracher we signed off the freehold equivalent of the croft to the Assynt Crofters. The owner of the estate that surrounded us was old school, with all the feudal baggage that comes with wanting to be master of all he surveyed, and there was no room in that agenda for a young couple hungry to do their own thing. We lived fully a mile and a half from his house, but for anyone who has come up against old money and old attitudes, you may understand how uncomfortable and uncompromising that can be. The Crofters of Assynt were reacting to the same feudal attitudes when they decided to pursue their purchase – and they had great lawyers! We couldn't afford lawyers, so on leaving we transferred the freehold equivalent of our croft to the Crofters so that any future Kerracher folk would have access to decent support should they too have to deal with more of the same. That particular facet of our life there was unpleasant; if you're not careful it's the sort of experience that can eat into you. It ate into us, for sure, but we found a way to turn negative feelings into positive action, and the experience colours our politics today.

Return to Kerracher

Rog – out there and fearless

Sutherland ranges – Ben Stack, Arkle and Foinaven

/ / / / /

Out from Kerracher and 11 kilometres east, at the head of Loch Glencoul, we find Glencoul Bothy empty, but not for long. Two Belgians, an American, a Brit and three Dutch appear later that evening and the next day. The lives of a couple of teachers, one of them a Shinto Buddhist, plus an author, a woodworker, a Steiner school teacher, a National Trust employee, a wind farm expert, a fish farmer and a public servant are illuminated by the dim glow of candles late into the evening. With whisky oiling our tongues, the chat across borders is convivial and wide-ranging – just how bothy life should be. Surrounded by each other's smelly wet gear, we find our mix of cultures weaves its way intriguingly through our conversation. There are new learnings, recognition of differences and of similarities too, shows of mutual respect, of shared concerns and shared food. All of us journeying for all sorts of reasons and all sorts of ends. Restless humans all.

- Route: Clashnessie to Glencoul bothy
- Distance: 22 km
- Weather: wet at the start but then calm enough, warm. Wind rising now, which is why we're hiding at Glencoul
- Wind: Green. Force 2/3 and rising from south-west
- Sea state: sheltered from the south and west today, so sea state low
- Events: Entering the old country; visiting Duart, passing Ardvar, visiting Kerracher
- Camping: at Glencoul; we use the bothy only for cooking and socialising.

Monday 25 July, Day 53: Glencoul, 0 km (weatherbound). Wind: Amber/Red.

Force 4 gusting 7, northerly. No kayaking today. Instead, Rog and I walk out beneath the Eas a' Chual Aluinn waterfall which, at 200 metres, is thought to be the highest free-falling water in Britain, then weave our way up the prominent nose of the Stac of Glencoul. From our high point, far off to the west, Kerracher is visible as a tiny white dot tucked vulnerably under the hill barely protecting it from the Atlantic looming vastly behind.

- · **Route: weatherbound at Glencoul**
- · **Distance: 0 km paddled**
- · **Weather: overcast but mostly dry**
- · **Wind: Amber/Red. Force 4 gusting 7**
- · **Sea state: white**
- · **Events: the climb, the bothy chat**
- · **Camping: outside the bothy at Glencoul.**

Tuesday 26 July, Day 54: Glencoul to Upper Badcall area 18 km. Wind: Green/Amber.

Late start to avoid the rain then into the wind on the nose – and it promptly rains anyway. Wind eases as we paddle back past Kylesku, and Loch a' Chàirn Bhàin is calm. However, a quite different sea meets us beyond Kerracher Point: lots of lump, and a wind still on the nose from the north-west. We are heading for Badcall and beyond, but as ever on this trip Mother Nature rules. I don't see much of the coastline as we head north, to be honest, so focused am I on simply staying upright. Rog, experienced kayaker that he is, is as unflappable as ever.

At the Badcall Islands we get some shelter, recover some composure, and then nose out between island and mainland to assess the going up to Handa. An even bigger and very gnarly-looking sea greets us at Farhead Point. Waves a good 2 metres, rising with the shallowing sea up around the islands and breaking with the bounce-back from the shore. I sit in the lump performing my usual assessment while Rog bravely heads out a little further into the liquid hills ahead. With my nerves already jangled after so many days out in tricky weather and especially after the recent headlands, I feel it would be just plain daft to cast myself into this sea at this stage of the trip and at this late time of the day. The wind is due to drop a notch or two overnight, so it's better to wait it out than let my ego rule the day.

We turn tail and find a wee cut in the coastline with just enough space to slip ashore on a rocky beach, and pitch the tents on a heathery slope to wait out the wind and the rains to come overnight.

Off Handa, balmy seas at last

Calm in Loch Laxford

Ardmore

The ancient rocks of northern
Sutherland ahead

- Route: Glencoul to Upper Badcall
- Distance: 18 km
- Weather: mixed: wind, rain, calm, sunny spells
- Wind: Green/Amber. Force 2 to 4 gusts
- Sea state: wind-blown and then calm in inner Loch Glencoul; out at sea, all is jumble and mess with, if we were to get too close to shore, clapotis and long waves bouncing back
- Events: lumpy water
- Camping: rough, steep slope, heather, rain, walk out to headland to check out tomorrow's sea.

Wednesday 27 July, Day 55: Upper Badcall to Ardmore, via Handa, 16 km. Wind: Green.

Out of our tiny cove and straight into some lump, heading for the sandstone island of Handa. But the sun is out, the wind is down and the wave height has dropped by a metre overnight. The finish lies within a day's paddle, but we decide to take two days over it, which will enable us to visit my old friends John and Marie-Christine Ridgway at Ardmore in Loch Laxford.

Handa, as ever in the sunshine, looks idyllic with its characteristic wedge shape and white beaches dotting its south and east sides. As we pull closer, the seas calm in the lee of the island and we pull into one of the beaches. Swim, yoga, taking it all in. I know Handa well, having kayaked around it countless times over the years, and I never tire of coming here. As a bird reserve it sits right up there with Noss up in Shetland, the Shiants of Harris and the Clo Mor cliffs up Durness way. At the thick end of the wedge, the Great Stack of Handa is where most of the action is, with thousands of guillemots, razorbills and fulmars, and many puffins too, all vying for space on the giant rocky outcrop. At the thin end, the terns, gulls, waders, bonxies and Arctic skuas hold sway. This year, however, there are additional visitors to Handa's shore – one-way visitors I'm afraid: dead gannets. These large seabirds don't nest at Handa, and while you see them hunting offshore it's always as a distant white flash. Yet today here they are, several of them, washed up dead on the beach. Bird flu is taking its toll of our seabirds this year, and the apparent rural idyll through which we are paddling is tainted with this daily loss.

As a counter to such depressing finds, the sun shines from a clear blue sky and as I'm performing some quiet yoga in a corner of this white sandy beach I feel the warmth just delicious on my skin. Days of such sunshine have been rare during my journey, so to have this weather at the closing stages is simply perfect.

We set off for Loch Laxford through more lumpy water north of Handa, then turn to shoot the gap at Sgeir Ruadh, and all is calm. Laxford, the name a Norse derivation of 'fjord of the salmon', is a place I hold dear. I first arrived here in the winter of 1985, aged just 20, to join a team of budding outdoor instructors at Ardmore, the outdoor adventure centre run by the indomitable John Ridgway. Fresh from Royal Marine training, I was at least fit and fired up, but I had much to learn about working with people and building skills for the job. John and his wife Marie-Christine gave me that opportunity.

As Rog and I paddle into inner Loch a Chadh-fi, the buildings of the tiny hamlet of Ardmore appear, and John and Marie-Christine ring the ship's bell up at the croft-house as we come into view. A warm welcome on a beautiful sunny day. Now in their early eighties and late seventies respectively, John and Marie-Christine, known to friends as MC, are hard to beat; an inspirational couple living inspirational lives, they have stories to tell that would curl the toes of even the most hardened adventurers. On top of that, their Adventure School gave clients and young instructors, of whom I was one many years ago, an opportunity to learn about themselves and others using the medium of the great outdoors. They ran this venture for 30 years, and as a result of people visiting from all backgrounds, creeds and geographies this tiny corner of the Highlands is bedded into the hearts and minds of thousands of people around the globe. Still today, almost 40 years on, I can quote John's three principles by which he ran his Adventure School and by which he expected all his working instructors to live: 'Positive thinking, self-reliance and leaving people and things better than you find them.' You can of course find more complex or nuanced principles wherever good people are running businesses or other organisations, but these three are simple to remember and easy to understand. And if you actively engage with them in your everyday life, they deliver on so many fronts.

John was his usual challenging self but no less welcoming for it and the conversation, as always with the Ridgways, was wide-ranging and stimulating. With the life experiences they have gathered, there is not much they've not already seen and heard, and as you're encountering minds as sharp as knives, you must be on your game to keep up with the fever of questions and counters as they come.

We stay over in the cottage next door, and the day here is sweet with memories. Ardmore was perhaps the start of the love affair with the Highlands that has coloured my life ever since. Having cut my teeth as a teenager with week-long solo camping trips into Rannoch Moor and other remote parts of Scotland, the years 1985 and 1986 at Ardmore, at the start of my twenties, cemented in me a powerful desire to make the Highlands my home. At the school there we ran adventure courses for companies, adults, young people, disadvantaged folk, wildlife enthusiasts and more, and taking direction from John I managed a team of instructors, built my people and outdoor skills, took care of maintenance, and even occasionally fed the salmon in the small farm in the loch. This was all formative stuff.

And as a backdrop there were always the hills, the mountains, the islands and the sea, slowly slipping into my very being, becoming something I couldn't do without, and creating some sort of early fulfilment. To run young fit and free across country on rough ground to Ardmore Point and to yell at the wind and spray stinging one's face as the sea spent itself on the rocks below – this was a freedom and a physical challenge I seemed to love. Mountain-running, ever since, has become an all-encompassing need, and in the intervening 37 years I have run Munros, mountain marathons, long-distance footpaths, the Great Wall of China and even the entire length of the Himalayas.

And the sea – the ever-present, ever-moving, ever-changing sea – this has been the foil for so much of life too. Making my living from the farming of salmon has taken me all over Scotland's seas, from Shetland to Argyll to the Outer Hebrides and everywhere in between. I know this place, and I feel deeply connected with its people and its politics too. The initial love of physical space has matured into an all-encompassing interest in our communities, both local and national, and over and above my work I've engaged with social and environmental politics as best I can over the years.

There is much happening in the Highlands, and those who move here purely for the view are missing what it is that makes this place such a stimulating and rewarding place to live. Many come north for a change of life. A few stay, and of those that do, the ones who contribute most are those who first learn to listen, to read, to seek to understand, and only then work out how best they might contribute. And if we are to continue to build the Highlands as a great place to live and work, it is contribution that is needed

I am clear, I hope, about the sort of Highlands I long to see. I hope for empowered local communities; people living *on* the land not just *off* the land; I want to see schools open and with enough children to offer a healthy social education; I want to see shops and pubs and ceilidhs and community woodlands and things for young people to do; I want to see a smaller percentage of holiday homes, and certainly few, if any, second homes; I want to see affordable housing, green energy, efficient use of our natural capital, communities vested in their area; I want to see incomers and I want to see those incomers keep quiet and listen before they believe they know how things work – and read some books (Jim Hunter would be a good start); I want to see us value 'local'; to think global and act local; I want to see great telecommunications and road networks, effective, well-maintained ferries and a planning system that is sensitive to the sense of community. And I want a brave government – braver than they are at present – who will pass laws to protect and promote Highland interest. Protect us from the seemingly unstoppable spread of the worst tourism has to offer; promote high-skilled businesses that provide jobs year round; protect our environment with a regulatory body that is efficient and enabling; and promote the devolving of power to a more local level, suitably supported with finance, whenever possible.

All this, and more, is indeed happening now – but it can feel a bit stalled on occasion. We have had a pandemic, an imposed Brexit and a Putin-inspired war with resulting fuel and inflationary crises, all of which are legitimate reasons for a slow transition from central to local. But we need to keep the pressure on, to keep volunteering, to step up and offer our skills and our time to the communities that form our lives.

And underpinning all that activity is a love of place. A spiritual quotient, if you will, to do well by others and by this place we all love.

For some this will be a feeling, for others a thought-piece, and for many it will be both. It is about harnessing our intellectual and our emotional intelligence quotients, our IQ and our EQ. It is ideas and people and geography and politics, and here I am at the end of my journey through this country I love, at the very spot where my own strong sense of place and my own intellectual and spiritual journey began.

/ / / / /

John, MC, Rog and I eat and chat until late in the evening, when the power cuts out, and then we chat some more, settling into the comforting dark of the Ridgways' front room here in this far north-western corner of the far north-western tip of Europe. Fanning out from this place are thousands of connections across the world: instructors, clients, businesses, schools. More than 30 years of inspiring the young and adults alike, the already motivated and the disaffected, the influential, and those without agenda. Here, all have been inspired to think a little differently, to toughen up both physically and mentally, to focus on working with others, to push boundaries and to try out new things without fear of failure. This is the Ardmore that sits in the collective memory of those touched by its magic over the years. And I am one of them.

And here we sit, Rog and I, with the two magicians who made it all happen. John and MC have lost none of their appetite for the new. Their daughter Rebecca and her partner Mark continue to run the Adventure School from the far side of the loch, and a new generation of aspirants is currently up on the slopes of the Foinaven Ridge, camping in the rain, carrying heavy packs and learning to 'stretch not cruise', before they too will head out into the world touched with a little of the Ardmore 'can do'.

Tomorrow is my final day. My two months are up and, while a part of me would be happy enough to keep going, up and round Cape Wrath, off to Orkney and beyond, that can all wait for another time, another adventure. For now, 56 days in, I'm happy to finish here at the most north-westerly settlement on mainland Britain and on the doorstep of Kerracher and Ardmore, where it all began. Tomorrow we head for Oldshore Beg Beach below Sheigra, then double back east to the fishing port of Kinlochbervie and journey's end.

- Route: Upper Badcall to Ardmore, via Handa
- Distance: 16 km
- Weather: light winds, some sun
- Wind: Green. Force 2. Light
- Sea state: lumpy either side of Handa, calm in Laxford
- Events: Handa, yoga, swim, dead gannets, lumpy seas, Ridgways
- Camping: croft-house at Ardmore.

Thursday 28 July, Day 56: Ardmore to Oldshore Beg and Kinlochbervie, 14 km. Wind: Green.

A beautiful day dawns. Breakfast with John and MC, and away on a sun-kissed ocean. Paddling out around Ardmore Point, we join a long rolling swell from the north-west. The sea is a mirror of the sky, the light a pale blue, fair weather clouds drift slowly overhead, and inland the mountains of Foinaven, Arkle and Ben Stack tower over the skyline. What a day to finish! No frazzle, no lumpy seas, no 'right on the edge' – just the promise of an end and the thoughts back over the past two months amongst the sea roads to the south.

For everything is south of me now, and as I paddle in to the finish, my mind casts back over the past 56 days. The quiet and gentle beauty of Argyll; the narrow channels with fast-running tides, gateways to the north at Dorus, Cuan, Kyle Rhea; the crossings too numerous to mention; the chances taken; the idyllic camping spots; the gnarly, rocky, really difficult camping spots; the exposure; dealing with fear; the relief at rounding a headland intact; the wind always plucking at my paddle and at my confidence; the birds, the seals, the otters, the dolphins; the people met, the kindness of strangers; the yoga on beaches and the being locked in the tent by rain; the long scary peninsulas, towering cliffs, no get-outs, the remembering to breathe; the sunrises and sunsets; the early starts and late finishes, the angst at the morrow's forecast; the rain, the wet kit, being wet most of the time, and later the realisation that I'm okay with it. But perhaps most of all, I will remember the solitude. The privilege to experience the utter quiet and a sense of humility, of oneness with nature, which has been forming even at this later stage in life. I've had the chance to think, to experiment with seeing, with noticing the small things, with introspecting and living, albeit for just a short while, a little differently from the everyday.

Underpinning it all is the journeying, and I have a wonderful sense of achieving a distance travelled. Conditions have been challenging, and mind, body and spirit have been tested for sure. This has been a journey both inside and out and I have work to do now to convert some of this thinking into doing.

/ / / / /

At Oldshore Beg beach, Rog and I swim and feed, then set off for the last time to paddle into the fishing village of Kinlochbervie, the last safe harbour before the remote bounds of Cape Wrath to the north. There, 38 years ago, in the hotel pub, rough and ready as it was, and marooned on the outer edge of Europe, a young impressionable 20-year-old, hungry for Highland life, watched open-mouthed as a grizzled old fisherman in filthy blue overalls and off his head with drink, crawled on his hands and knees out of the ladies' toilet. There be stories galore for another time.

- · Route: Ardmore to Kinlochbervie
- · Distance: 14 km
- · Weather: gorgeous, blue, warm breeze
- · Wind: Green. Force 2. Light
- · Sea state: long low swell
- · Events: sun-kissed final day at last, Oldshore Beg beach, the finish.

Lost in the blue

A man and his boat

Kinlochbervie and journey's end

Tired, happy and 6 kgs lighter

Postscript: Reverence

I hope my musings will excite your interest in the natural world. I hope they will open your eyes to the people and animals who live here in the Highlands and make it what it is, and most of all I hope it will encourage you and others to take an interest in the changes that surround us and do something to help.

Change – ecological, social and political – has always been with us, but there is a feeling, underpinned by good science, that negative ecological change is speeding up. For the river kayaker fast approaching a potentially deadly set of rapids, the river should run slow in your mind, emotions and route choice under control. But if it's running at fast forward, if there is panic, then that's a sign that control is lost and the risk of drowning is real. Our climate change feels like the latter to me. It's in fast forward, we have barely any control and we're careering towards that deadly rapid. More droughts, more floods, more storms, more wildfires, more climate refugees, more political unrest, increasing uncertainty. And still, economic growth is the solution for human prosperity peddled out by political leaders worldwide.

There are other leaders out there – thought leaders with clear heads on what we need to do to reverse the climate catastrophe already under way. But they are too often marginalised and not enough valued, wedded as we seem to be to an economics without environmental conscience.

It can feel hopeless, but we can't afford to be without hope. That just leads to apathy, inaction and the continuation of the reckless behaviour that has brought us here. Instead, we need to listen again to what the animals and plants surrounding us have to tell us. We need to rediscover enough humility to recognise that we're part of nature and not separate from it. And we need to understand and then accept our impact, and use our imagination to fashion a vision for our planet where humans, animals and plants can live in better balance.

Imagine a world without the seabird's cry, without the piping call of the oystercatcher, the surreal drumming of the snipe, or the pealing call of the eagle high above. Imagine a sea devoid of fish, woodlands without birdsong, and an empty sky. If we are indeed heading that way, then it's partly because we don't 'see' or feel the wild things, the fellow souls that surround us. And because we don't see them, we are losing our ability to care. The biological world is fast becoming something we know only in theory, something we see only in two dimensions on our tiny screens – and even then it's invariably the distressed polar bear, far away and too distant for most of us to feel a genuine connection. And by not connecting directly, we're losing the feeling of it, I fear – we're losing the wonder and thus the reverence we need to act.

My own sense of wonder, of awe, has been rekindled by spending time bedded in nature. Feeling small, humbled and exposed, I have learned anew how to listen and observe, and how to see myself as part of nature and not separate from it. I have watched animals and birds go about their lives despite the doom and gloom of the climate crises. Above and below the surface of human-induced plastic pollution, warming seas and overfishing, they continue to rear their young, to feed, socialise, migrate – and that gives me hope. Hope that all is not lost, that, it is not too late, and that if we can slow and then reverse the ecological degradation we are causing, then life will find a way.

But to make that happen we have to care enough to change our ways. And caring will only come when we're engaged with the emotion of it, when we're humble enough to see ourselves as part of the whole, and when, like the Native American, we can demonstrate some reverence. If we can finally come to see ourselves as much affected by our actions as are those fellow wild souls around us, then we might start to care enough to change our ways.

I hope my journey through the Highlands and associated musings may not only strike a note of hope but also act as a call to action. I hope I have provided a reminder that the natural world, vast and resilient, is still out there beyond our mobile phones and urban lives, and I hope I have helped make that world, both human and wildlife, a little more visible. I hope too that I have encouraged us all to be more reverential of the environment that supports us, and to add more value to it than we have done to date. If we can embed care for our environment into our sense of community as a core spiritual value, as central to our identity, then our economic and political decisions will follow.

All is not lost; we just need to add a little reverence and then, perhaps using the ideas outlined in Day 48 or those in the notes below … act.

Bird flu

Sedum, tiny life finding a way

The colour beneath

Sundew - midge eaters?

Kelp

Lichen

Maerl cast up

Heath spotted orchid

Rock art

A splash of ivy colour

Sea urchin – out of his depth

Seaweed – nature's art

Molluscs – shell beach in the making

Carpets of Thyme

Addenda

How to act, then?

Here are some ideas.

I have no wish to be prescriptive here, as each of us has our own climate-specific concerns and preferences. It's also all too easy to become overwhelmed by 'to do' lists, and I recognise too that climate anxiety can be a paralysing issue for some. So I'm not going to provide a 'must do' list, and I'm certainly not going to prescribe any specific actions.

Instead, I'm going to offer up some broad ideas on how to think about taking action and include website links to just a few of what I think are fantastic organisations, providing simple and practical guidance on how we, as individuals, can mitigate climate change and protect the natural world.

You're almost certainly doing some or perhaps all of these things already:

Commitment: firstly, decide for yourself that this is important enough for you to take action … and then commit to the task. Don't focus too much on outcomes but instead focus on creating new habits for yourself, new instincts, so that your environmental ambitions/concerns become central to the decisions you make during your daily living.

Food: the food we eat comes with a carbon footprint, and not all foods are equal. Consider eating more veg, eating more seasonally and purchasing more locally.

Waste: food waste accounts for around 8 per cent of global greenhouse gas emissions, so work at throwing away less food. Packaging helps prevent much food waste but it is also a well-known polluter, so do away with using and purchasing single-use stuff and move towards using multiple use containers. 'Reduce, reuse, recycle' is a well-known and powerful call to arms.

Travel: two wheels good, four wheels bad … and wings are worst of all. Minimise use of the most polluting forms of travel and if you can't avoid it, contribute to an offsetting scheme. Food travels too, so check the packet before you buy and if it's travelled half-way round the world, consider a local alternative.

Keeping warm: heating our homes is a major contributor to global warming, and no-one likes being cold after all. Insulation, upgrading inefficient boilers, turning off unnecessary radiators, turning down the thermostat a notch or two, putting on another layer of clothing – all these will reduce your impact. I swear by thermal underwear … but will never be a fashion guru as a result!

Volunteering: giving your time and expertise to a worthy charity, NGO or social enterprise etc. can make the difference between that organisation existing or not.

Giving money: if you are short on time and long on cash … then give money. Do your homework first so you're comfortable that your hard-earned cash will be used well.

Investing: investigate whether or not your pension pot or other investments are holding shares in companies or industries that don't fit your values. If they are, change them.

Advocating: there are all sorts of opportunities here:

- If you're in business then investigating how your company might become a B Corp organisation would be a clear public statement of intent. See https://bcorporation.uk/b-corp-certification/
- As a consumer, as well as your purchasing power, you also have the opportunity to question businesses directly. Retailers, especially the big supermarkets and clothing stores, for instance, are increasingly sensitive to their customers' values and concerns. You could contact them, politely and having done some research

in advance, to ask them what are their policies on the areas of your interest.

- As a citizen, you can advocate both locally and nationally by engaging with your local community and/or district councils and with your MSP or MP. Again, I have found a polite and well-researched approach always works best, and if there is more than one of you, i.e. if you make your approach as a group, then that will help too.

And finally, **communicating**: not everyone likes to share their thoughts and actions, but with the realities of global warming becoming increasingly manifest in our daily lives, the time for keeping quiet is over. Not all of us can be as awesome a communicator as Greta Thunberg, but we can at least talk with our friends, our colleagues and with organisations if we so wish.

So be brave … advocate.

And finally for this section, I did promise not to provide a long list of websites, and I won't. Here are a select few that might help you on your way:

B Corp: https://bcorporation.uk/b-corp-certification/ This business certification organisation assesses business practices and outputs across five categories: governance, workers, community, the environment and customers.

The Grantham Institute: https://www.imperial.ac.uk/stories/climate-action/ The institute offers '9 things you can do about climate change' and '9 things you can do to protect the natural world'. You'll find lots of additional links and ideas here to help guide your own environmental mitigation journey.

Carbon Savvy: https://carbonsavvy.uk/action/ A community interest company working to support individuals, businesses and councils get to net zero by offering advice and footprint calculators.

Scottish Communities Climate Action Network: https://sccan.scot/about/ A communities-focused volunteer-led network connecting individuals into local networks and initiatives.

Transition Black Isle: https://www.transitionblackisle.org/index.asp A terrific example of the Transition Network Communities movement, https://transitionnetwork.org/ . I've provided the link to my local Transition movement here north of Inverness in Scotland so you can get a feel for how these local people convert the ideas from the central network into reality on the ground.

I hope that little lot helps. I personally am acting on several things in the list above but not all, and I'm working on the gaps. Good luck on your own journey.

Carpets of Thyme

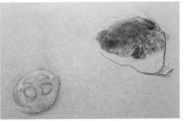

High and dry

Strandings

Sunset

Forecasting the weather – a kayaker's-eye view

Yes, today you can open up your mobile and get an hour-by-hour forecast for just about anywhere in the world. But in that instant fix there is little if any long-term learning. To complement the technology it's a good idea to build up, over time, an ability to glance at the sea and the clouds and develop a more general forecasting feel. I use the scientific forecast service, too, of course and have a personal preference for the Met Office's slightly broader-brush approach. Other forecasting services are available.

Key for the kayaker is the wind speed and direction. That determines the sea state, and after a while you can get pretty good at forecasting what the sea will be like out of sight round that headland or in the forecast calm after a previous day's blow. The official forecast must always be interpreted for the local conditions in time and space. Do that, and you'll have a nice life out on the water.

I use the Beaufort scale, which splits the wind speeds into twelve, from calm to hurricane. Calm is Wind Force 0 and hurricane is Wind Force 12. A gale is Force 8.

For sea kayakers, all is well until you start to hit Force 4 to 5. In those winds on an exposed shore the sea can be messy and confused. Where the wind is blowing off-shore all might still be calm, but if you hit a gap between islands or a downdraft from the mountains, watch out; if you capsize you'll be blown out to sea.

There are so many variables when making decisions about when and where to paddle, and while the books and websites will tell you a lot you can't beat a bit of experience through time spent on the water.

Hopefully you'll get it right, but no doubt sometimes you'll get it wrong ... and goodness knows I've done plenty of the latter over the years!

For this journey, rather than use the Beaufort scale alone to describe the winds, I have converted it to a 'green, amber, red' system, which I hope will help the non-seafaring reader identify more easily with the feel of each day. Green is anything from calm to gentle breeze (Force 0 to 3); amber is moderate to strong breeze (Force 4 to 6); and red is anything from near gale (Force 7) upwards.

If the wind has blown strongly for a day, the following day may be forecast as calm but the sea state can still be lumpy. Green does not therefore necessarily mean it's an easy day on the water. Amber means it's a lumpy day on the water for sure. And if I'm out on the water in the red zone, then I'm either feeling a bit devil-may-care or I've made a poor decision!

Beaufort wind scale Force	Mean wind speed (in knots)	Wind descriptive terms	Probable maximum wave height out at sea (in metres)	Ed's Green, Amber, Red system
	A sea kayaker may pootle along at 3.5 knots, step it up to 4.5 knots for an exposed cross-ing, and use a 'get the hell out of here' rate of 5.5 knots to, well, get the hell out of there.		Very dependent on local con-ditions when coastal paddling	
0	0	Calm	-	Green
1	2	Light air	0.1	Green
2	5	Light breeze	0.3	Green
3	9	Gentle breeze	1.0	Green
4	13	Moderate breeze	1.5	Amber
5	19	Fresh breeze	2.5	Amber
6	24	Strong breeze	4.0	Amber
7	30	Near gale	5.5	Red
8	37	Gale	7.5	Red
9	44	Strong gale	10.0	Red
10	52	Storm	12.5	Red
11	60	Violent storm	16.0	Red
12		Hurricane		Red

Full route details

Key
n = necky paddles; b = bothy; w = weathered off (whole or major part of the day); f = ferry; p = portage; b&b = b&b

Day	Date	Route	Distance	Wind force	G/A/R	Other info	Jelly baby day
1	03/06/22	Gourock to Ardyne Point	21	3 to 4	amber	n	✓
2	04/06/22	Ardyne Point to Port Leathan (east side Loch Fyne)	29	2 to 3	green		✓
3	05/06/22	Port Leathan to Gigha	29	3	green	n, p	✓
4	06/06/22	Gigha to Stotfield Bay	25	1	green	n	✓
5	07/06/22	Stotfield Bay to Craignish	37	2	green		
6	08/06/22	Craignish to Luing	20	2 to 4	green/amber	n	✓
7	09/06/22	Luing to Oban	23	3 to 4	green/amber	n, b&b	✓
8	10/06/22	Ferry to Craignure then 2 km north of Craignure	2	5 to 7	red	f	
9	11/06/22	Weatherbound 2 km north of Craignure	0	6 to 7	amber/red	w	
10	12/06/22	Craignure to Calve Island (east of Tobermory)	25	4 to 6	amber	w	✓
11	13/06/22	Calve Island to Salen, then lift to Mallaig	20	3 to 6	amber	n, b&b	
12	14/06/22	Mallaig to Sourlies (Loch Nevis)	23	3 to 4	green/amber	b	
13	15/06/22	Weatherbound at Sourlies	0	6+	red	b, w	
14	16/06/22	Weatherbound at Sourlies	0	6+	red	b, w	
15	17/06/22	Sourlies to west of Ardintigh Bay	12	5 to 7	amber/red	n	✓
16	18/06/22	Ardintigh bay to Eilean Giubhais (south side Loch Nevis)	8	5 to 7	amber/red	n, w	✓
17	19/06/22	Eilean Giubhais to Dun Ban Bay (south of Doune)	8	3 to 6	green/amber	n, w	✓
18	20/06/22	Dun Ban Bay to 3 km west of Skye Bridge	37	2 to 4	green/amber		
19	21/06/22	West of Skye Bridge to Portree	37	2	green	b&b	✓
20	22/06/22	Portree to Invertote	19	4 to 6	amber	n	✓
21	23/06/22	Invertote to Port Gobhlaig	21	4 to 6+	amber/red	n	